tITLE:

PHI CODE 108

PHI CODE 108

bY

JAIN 108

the
108
Code

In Adoration + Receivership
of the Phi 108 Code

(tHE bOOK oF pHI VOLUME tHREE)

sUB- tITLE:

tHE 108 cODES, aN iNTRODUCTION

eDITION 2

aUTHOR:

jAIN 108

yEAR:

2010 Jain ©

iSBN: 978-0-9757484-2-8

Phi Code 108

P
H
I

C
O
D
E

1
0
8

eDITION 2

(suggesting that Editions 3 & 4, will be updated versions of this book, as it forever evolves, will be formatted in the **Golden Rectangle** size 210 mm by 210xPhi mm, and will be typed in the highly professional **INDESIGN**, the assignment delegated to my trustworthy Friends of Growth, so that I can get on with the essentially work of Phi Dissemination and Recrudescence.

www.jain108.com

Saint Francis of Assissi wrote:

"The Result of Prayer is Life.
Prayer Irrigates the
Earth & Heart"

(for a creative poem written only using the letters from Jain's **EARTH-HEART** go to this link:
http://www.jainmathemagics.com/product/88/default.asp
or
view it in full in the Appendix on page 184 at the end of this book.

cONTENTS
tHE bOOK oF pHI, Volume 3,
subtitled: The 108 Codes, an Introduction.

nb: (Most words highlighted in the text of this book
    are listed alphabetically in the Index).

the 108 PHI CODE TRANSMISSIONS

by JAIN 108

PHIomETRY & MathLetics:

It so happen that I drew the short straw
for the release of the Phi 108 Codes.
It comes at a time when humanity needs
to be introduced to this lost topic that I call **Phi-Ometry**.

Understanding the essence of the Phi Mysteries, is an important
subject in Jain's University of Sacred Geometry.
A visual understanding of Nature's Living Curvation will prompt the
student to learn Rapid Mental Calculation, towards the age coming
where every child in the future will be doing maths in their head, as
we were suppose to from the beginning before we became dumbed
down from overly processed foods and poisoned waters.
Jain Mathemagics will give your child the confidence and exalted
memory power to perform speed mathematics. In a sense, it is a
form of **Math-letics**, as in the "Athletics of Mathematics".

by Jain, dec 2007, Mullumbimby Creek, far north NSW

****

"SACRED RELIC 117"

If you gave me a sacred relic
    from Tibet, I would walk away
If you gave me a sacred relic
    from the Vatican, I would not accept it
If you gave me a sacred relic
    from Mecca, I would remain aloof
If you gave me a sacred relic
    from Uluru, I would purse my lips
...
though,
if you gave me some
ancient timeless eternal infinite mathematical palindromic-phi-pi
gem pattern like Sri 117
I would accept it in my Heart

Poem by Jain 108 xmas, 2007, Mullumbimby Creek, NSW, OZ

(acknowledging the Phi Code pattern of 12 x 9 which = 108,
where all the pairs sum to 9,

1	1	2	3	5	8	4	3	7	1	8	9
8	8	7	6	4	1	5	6	2	8	1	9

but there exists a final pair that is a double pair of 9s, thus the flip
side of the 108 code can be viewed also as 108 + 9 which = 117 !)

definition of "CODE"

"...the term 'Code' has a very broad meaning in everyday
language, and it is often used to describe any method for
communicating in secret..."

[quote from "The Code Book" by Simon Singh, pub. By Doubleday,
New York, 1999]

Grokking The Phi Codes...

By grokking the meaning of these Phi Codes, the reader will evolve
to a point where they can not even say: "I send you Love and
Light" because this becomes a limiting belief, that implies that
there is something inherently wrong with that person. The reason
we begin to think in Unity Consciousness is because the Phi Codes
slay the Victim Mentality, encourages Unconditional Love, assists in
staying in The Now and therefore brings true Peace.
"Luceat Lux Vestra" = Let your Light Shine.
It is my path to make these Phi Codes, these Liquid Light
Transmissions and Transmutations simple and easy to understand.
No point in talking a lot about "e" known in mathematics as the
ellipicity of a spheroid, who would be interested save a handfull of
nerds.
No point in focussing on the **pHARMacratic tyrannies** and the
world domination of the clandestine Corporatocracy, because what
we focus on or yell about, we become, we magnify that anger,
better to remain heart centred and become the change, become the
new model, become a stellar nursery of infinite possibilities.
What I mean by this, is that if you truly want to create peace, then
you have to be peaceful. You know this is true, that the obvious
way to cease war is to stop the war within yourself, to change your
own mindset, and start from within your own self. Simply, it means
that you must do personal disarmament. All this time you thought
you had to battle against the **corporatocracy**, but since you
imbued Phi and are now Engrailed, the work is within, there is no
more mud-slinging or blame game. You begin to become self-
organized.
When you embrace your natural state, then the external atrocities
no longer affect you, no more reaction to bio-terrorism, who cares
about the cabal of inter-breeding families seeking to impose a
global fascist dictatorship of total human control. The peace in your
heart will become the global politics.

Jain's Dyslexic's Spelling: "SPONE"

Here is an account/story on how dyslexic I become when my brain is firing fast, not really concentrating, being in the future
eg: I am typing speedily on my computer, the sentence:
"In biblical Babylon, the world spoke one language".
Whilst halfway through, writing the word "spoke", I already have mentally formed the next words to write which is "one language". So instead of writing "spoke" I mindlessly and unawarely write "spone" which appears to be a fusion of the two words "spoke & one".
Actually it's a good word. It's a very efficient mistake or natural compounding of words. The brain actually likes the Path of Least Resistance, it's one of the Law of the Universes, like the **Law of One**, known as the Law of Economy. We all love new inventions and seek time saving devices. Actually, dyslexia is a gift of genius, as it unifies, merges, sorts and recreates data to fit into it's package of the ever evolving english language:
"In biblical Babylon, the world spone language".
I like it.
Thus talking about **One Language**, the point is that what the world needs now is a **One Mathematics**. With the global internet and international currencies, we can see that the numbers zero to nine are the language that we are all communicating with on earth, as with all telephone numbers, credit card transactions, home addresses, reading the time or clock, it's all numbers in Base Ten, how welcomed it is, pushing us all towards a NEO global economy **ONEarth**.

~ Jain 108 ~

QUOTES from Others:

"omnia apud me mathematica fiunt:
With me everything turns into mathematics."
~ **Descartes** ~

"Wherever there is number, there is beauty."
~**Proclus** ~ (410-485 A.D.)

"Without tradition,
art is a flock of sheep
without a shepherd.
Without innovation,
it is a corpse".

~ **Winston Churchill** ~

"You use a glass mirror to see your face:
your use works of art to see your soul".
~ **George Bernard Shaw** ~

You use Sacred Geometry to remember who Your Are;
You use Mathemagics to understand Fixed Eternal Design
& the Divine Symmetry and Living Curvation of your Soul.

~ **Jain 108** ~

"If U want 2 C something
Open your Eyes

If U want Know something
Close your Eyes"

~ **Ramtha** ~

I AM A
SELF-SUSTAINING
SELF-AWARE BEING,
WHOSE FRACTALITY,
INSIDE AND OUT,
IS AS INFINITE
AS THE PERFECT
SHAPE OF GOLD...

..."we need to become like pine cones"...

(~ **Dan Winter** ~ 2009
www.goldeninfo.com)

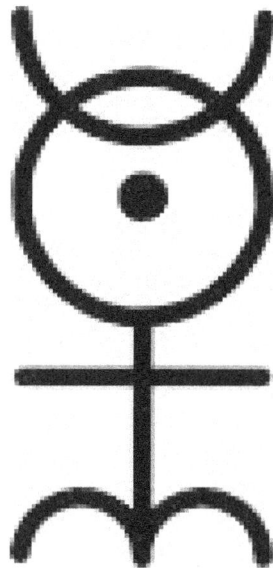

John Dee's glyph, whose meaning he explained in **Monas Hieroglyphica**. The Hieroglyphic embodies Dee's vision of the Unity of the Cosmos and is a composite of various esoteric and astrological symbols.

A Tribute to JOHN DEE

QUESTION: what did Jain mean, when he talked about his " **PHI INCUNABULA**" ???

ANSWER: not sure, but I found this page of information from Vincent Bridges book, that defines the actual meaning of "**Incunabulum**" which translates to **Jain's rare library of phi** and magic squares and rapid mental calculation etc.... Here is the definition from John Dee, the dude often remembered as Queen Elizabeth's personal astrology or court adviser...
The following quote is a good reference to numbers extracted from Vincent Bridges book.

"...John Dee was an intensely pious Christian, but his Christianity was deeply influenced by the Hermetic & Platonic-Pythagorean doctrines that were pervasive in the Renaissance. **He believed that Numbers were the basis of all things and the key to knowledge, that God's creation was an act of numbering**. From Hermeticism, he drew the belief that man had the potential for divine power, and **he believed this divine power could be exercised thru mathematics**. His cabalistic angel magic (which was heavily numerological) and his work on practical mathematics (navigation, for example) were simply the exalted and mundane ends of the same spectrum. His ultimate goal was to bring forth a unified world religion through the healing of the breach of the Catholic and Protestant churches (at war with each other at the time, note the Thirty Years War) and the recapture of the pure theology of the ancients.
According to scholars, in his lifetime Dee amassed the largest library in England and one of the largest in Europe. The word "**incunabulum**" means cocoon or cradle and refers to something either in infancy or in metamorphosis. It is also referred to as the printed books prior to 1501, and has been extended **to mean any rare and hermetic collection of books**. John Dee used the term incunabula to refer to his library, which contained the most comprehensive collection of information since the Library of Alexander burned.

Question:
What do you mean when you say that Jain's Phi Code 108 is an Entelechtic approach to learning?

Answer:
Entelechy is defined in the medical dictionary as:
"...Completion: full development or realization; the complete expression of some function. A vital principle operating in living creatures as a directive principle..."
Thus Jain's curriculum: "**The Art Of Numbers**" or "The Translation Of The Fibonacci Numbers Into Art" is an entelechtic or wholistic approach to educating the student in the vast realms of fractal mathematics, mainly because it is concerned with the total human, the highest form of learning, the truth amidst the entangled web of theories. Because it is revelatory, it also addresses or impacts the mind, the emotions and the phi-based etheric fields that govern the functioning and well-being of our Spirit-Mind totality. The entelechtic Art Of Number therefore shapes the student's porous Higher Mind into a Funnel that downloads the essence of timeless, eternal Mathematical Truth. This is the 108 Code, hidden in the geometry of flowers!
Whereas orthodox mathematics as taught in most schools: the factory-style curriculum that is bland and tasteless and soulless, has lost sight of the original higher principle or directive that exalts Vortex Mathematics of Nature as a Star Language or Angelic Script of Vibrational Laws no different to the lost knowledge of the Geometrical Angles of the Angels that became known as Anglish and now known as English.
Thus Jain's "Art Of Number" that is rewriting all of Einstein's clumsy equations in precise terms of the Phi Ratio 1:1.618033... teaches the receptive student "The Language of Universal Pattern Recognition" that makes the Invisible to become Visible.

By Jain,
written at **12:34**pm + **56** seconds on the **7**th of the **8**th of 0**9**.

(Art
by

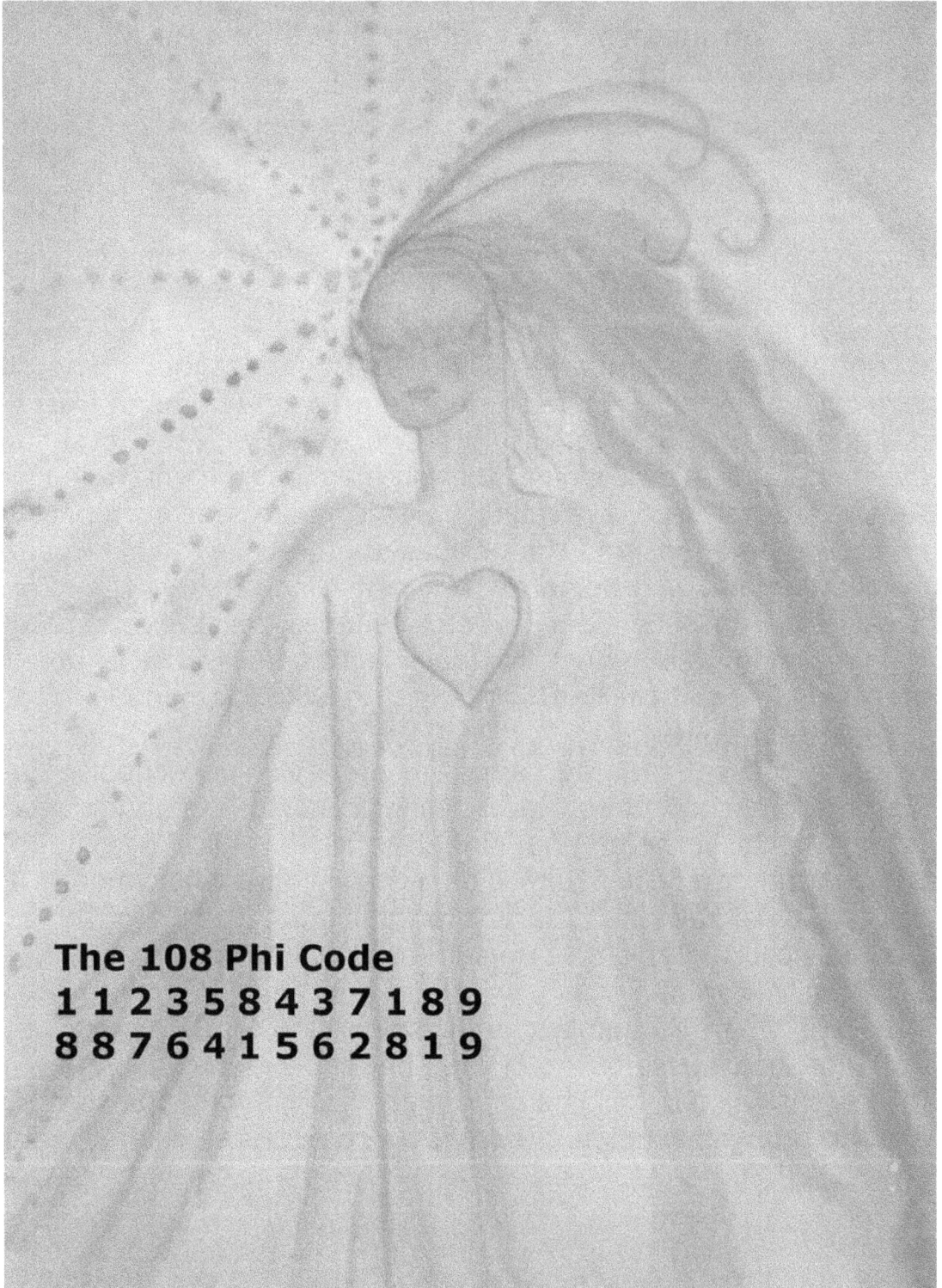

The 108 Phi Code
1 1 2 3 5 8 4 3 7 1 8 9
8 8 7 6 4 1 5 6 2 8 1 9

Jain, 1994 "Angelic Being releasing the first of two Phi Codes")

lOST sECRETS oF tHE pHI cODE

(ARTICLE for Magazines: April 2006)

pART 1

**WHAT YOU ALREADY KNOW
ABOUT PHI and The FIBONACCI SEQUENCE**

pART 2

**WHAT YOU DON'T KNOW
ABOUT PHI and The FIBONACCI SEQUENCE:**

JAIN'S DISCOVERY of The REPEATING 24 PATTERN

pART 3

**WHAT YOU NEED TO KNOW
ABOUT PHI and The FIBONACCI SEQUENCE**

**TIME CODE 12:24:60 ENCRYPTED
in The FIBONACCI SEQUENCE and PASCAL'S TRIANGLE**

pART 1

WHAT YOU ALREADY KNOW
ABOUT PHI and The FIBONACCI SEQUENCE

Most people are now aware of the importance of the Fibonacci Sequence, thanks to the best selling book: "The Da Vinci Code", by Dan Brown, whether or not the historical facts are true or false is not as important as how far-reaching the effect this book has had on the mass-consciousness. The success of this book means that a few more million people have heard of the once well-known Fibonacci Sequence of numbers:

0, 1, 1, 2, 3, 5, 8, 13, 21, 34, 55, 89, 144 etc

They are derived by having a starting point of 0 and 1, the substance and binary language of all computers, and adding these two beginning numbers together:

0 + 1 = 1

This gives the third number of the sequence: 0, 1, **1**,

Then add the previous "1" to the last "1" which gives: 1 + 1 = 2

Giving the fourth number of the sequence: 0, 1, 1, **2**,

It's like adding the Past, to the Present, to give the Future, a veritable Trinity of Numbers, thus the next number would be:

1 + 2 = 3 and the continuing sequence becomes 0, 1, 1, 2, **3**, and so on.

When we decimalize these ratios, they approach the number: 1.618033988 and travels to infinity. We call this relationship, when indexed against Unity, as the Phi Ratio: **1:1.618033988** and symbolize it by using a Greek letter of the alphabet that gives the "f" sound, called "**Phi**" or "**f**" (whereas most people are familiar with the "p" sound in the Greek language, called "Pi" or "p"). Some people incorrectly pronounce "Phi" as "fee" but I write it here to get it right for the future generations of **Fibonatics** that the correct pronunciation is expressed as the same sound heard in the nursery rhyme:

"Fee, **PHI**, Fo, Fum, I smell the blood of an Englishmun".

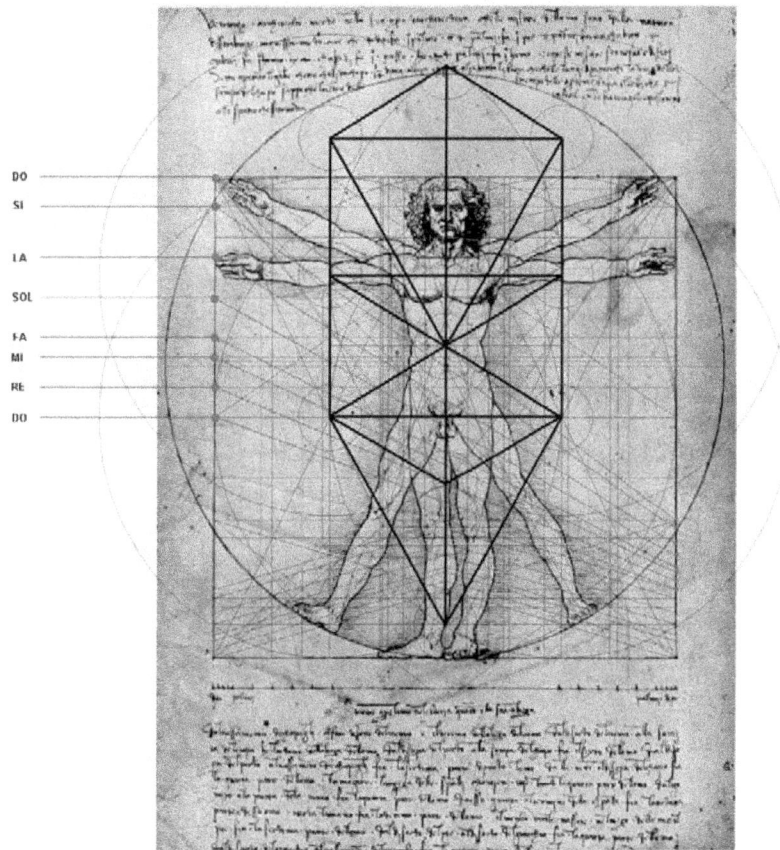

A IMMAGINE E SOMIGLIANZA

Connie Johnston © 2002 © Alfonso Rubino

Fig 1
**The Human Body is the ultimate expression of the Phi Ratio.
Where the elbow bends,
compared to the whole length of the human arm,
is in the Divine Proportion
expressed mathematically as 21:34 of the Fibonacci
Sequence. Classical biometrical Image by Leonardo Da Vinci.**

The Phi Ratio is the living mathematics of Nature. We see these Fibonacci Numbers in the pentacle form of many flowers like the passionfruit; we see it in the Pine Cone, 8 spirals going one way intercepted by 13 spirals going the other way, and in the Sunflower, we see 21 Clockwise spirals versus 34 anti-Clockwise spirals.

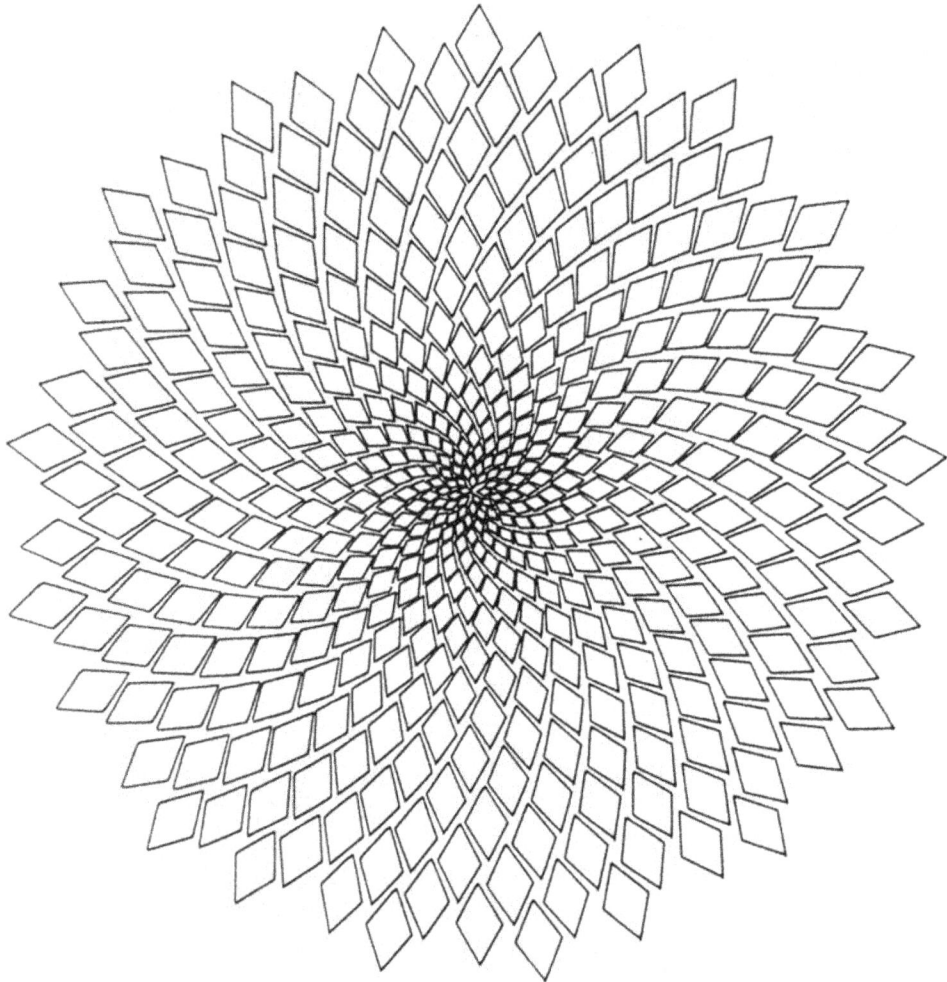

Fig 2

**Computer Rendition of the Sunflower Floret
showing 2 distinct spirals:
34 Clockwise and 21 Anti-Clockwise.**

We also see these numbers as the approximate distances of the planets from the sun, known in the scientific world as Bode's Law. Why are we awe-struck when we view **Sacred Architecture** like the Parthenon of Greece and the Pyramids of Gizeh. It has to do with Shape: the façade of the Parthenon is a Golden Rectangle, in the proportion of 21:34, which is identical to the proportion of the Human Body like where the elbow bends, where the knee bends, thus temples are reflecting our own internal symmetry, so we are literally seeing ourselves when we view beautiful buildings constructed in this awareness.

So for thousands of years, great artists used this Fibonacci Sequence as the blueprint for their canvass. That is why we are attracted to the proportions of the famous **Mona Lisa,** as the outer rectangle that contains her form has the blueprint of the Golden or Phi Rectangle in the ratio of 21:34. In fact, she has become the global embodiment of the Phi Ratio.

When I was in Tibet in 1990 I was with an artist who was painting his Thangkas or holy images of Buddha, and noticed the book that was the basis of all his measurements. It was referenced with the Fibonacci Numbers, making it law that every Buddha must be drawn in the Phi Rectangle otherwise it was considered bad energy. The classical book on this subject that introduces you to the universality of the Golden Mean as found in Nature is:
"The Power of Limit's" by Gyorgy Doczi, subtitled: Proportional Harmonies in Nature, Art and Architecture, an example of the Buddhist's canon is shown below.

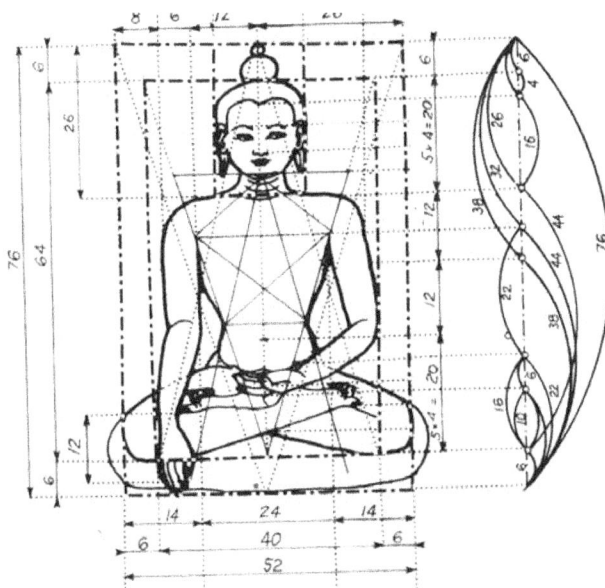

Fig 3
3 Golden Rectangles in the Biometrics
on Buddha's proportions.

Here is a very powerful or psycho-active diagram that changes consciousness, a diagram that brings together the worldview of the Scientist (factual measurements of the helical DNA) and the Metaphysician (someone who intuitively understands the higher meaning of the Phi Ratio). This bridging of 2 worlds is part of the change going on the planet now. When we view the twisting DNA molecule as it fit's inside of a cylinder, and measure the critical distance from one atom to align itself to it's original position after one complete turn or rotation around the central axis, like a spiraling vine would, our dear scientists noticed that the diameter was exactly 20 AU (Angstrom Unit's = one ten-millionth of a metre) and the vertical axial distance was 34 AU. You will notice that this relationship of 20:34 is almost identical to the Phi Ratio of 21:34. Metaphysicians are intrigued by this DNA Phyllotaxis and the mathematics of DNA approximating the Phi Ratio. They conjecture that perhaps we are not currently in the Phi Ratio, but are evolving to that ideal state. (A more paranoid interpretation would be that Someone, or some Race, in our distant past, created us humans as slaves in goldmines, engineered by the Extra-Terrestrial Masters of Gene-Splicing who were harvesting our planet for our DNA, and had genetically engineered as thus). This is one purpose of Sacred Geometry, is to start asking questions that bridge beliefs, knowledge and history of who we are.

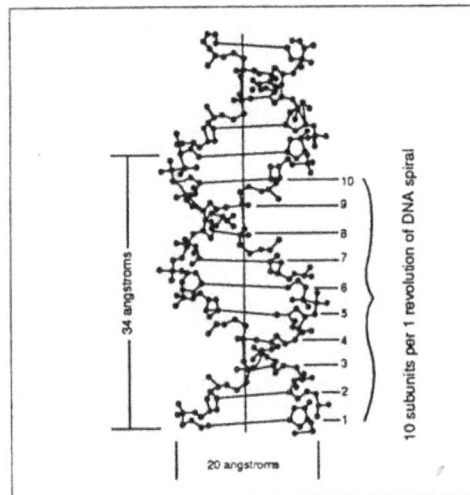

One 360 degree turn of DNA measures 34 angstroms in the direction of the axis. The width of the molecule is 20 angstroms, to the nearest angstrom. These lengths, 34:20, are in the ratio of the golden mean, within thelimits of the accuracy of the measurements. Each DNA strand contains periodically recurring phosphate and sugar subunits. There are 10 such phosphate-sugar groups in each full 360 degree revolution of the DNA spiral. Thus the amount of rotation of each of these subunits around the DNA cylinder is 360 degrees divided by 10, or 36 degrees. This is exactly half the pentagon rotation, showing a close relation of the DNA sub-unit to the golden mean.

Fig 4
**DNA Molecule, as it turns helically,
exhibit's approximately the 21:34 Phi Ratio**

When we convert these Fibonacci Numbers into 2-Dimensional squares, they fit snuggly into the Golden Rectangle whose subsequent quarter-circle arcs form the familiar Golden Mean Spiral:

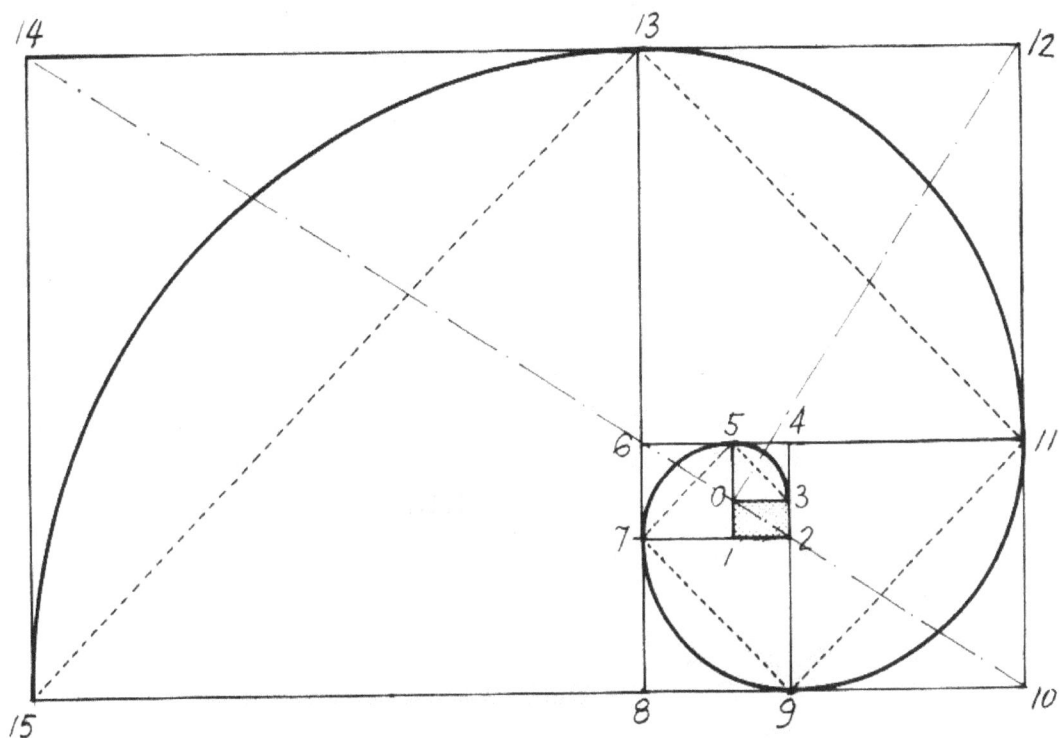

Logarithmic spiral, typical of shell growth. Successive stages of growth are marked by "whirling squares" and golden rectangles growing in harmonic progression from center **O** outward.

Fig 5
The Fibonacci Numbers can be represented as Squares.
Each square produces a quarter-arc circle.
The overall bounding shape that contains the
Golden Mean Spiral is the Golden Phi Rectangle.

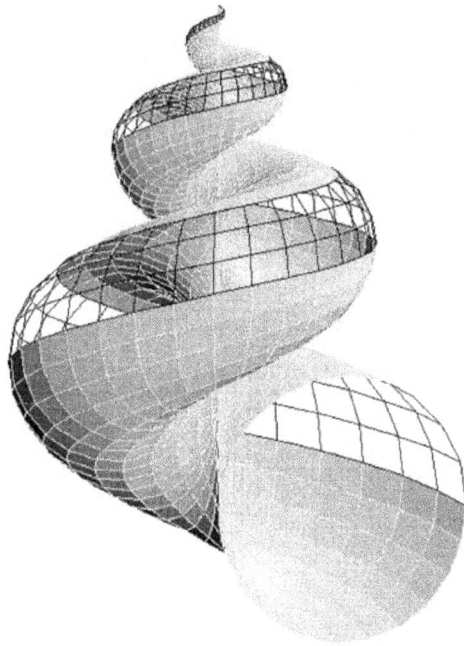

Fig 6
**If you were to visualize each flat square
of the Fibonacci Numbers as a cube,
you would produce a 3-Dimensional form
of the 2-Dimensional flat spiral.**

Fig 5a
Horn of Goat showing the 3D Phi Spiral later demonized.

Fig 7
**When we convert these Fibonacci Numbers
into 3-Dimensional cubes,
the resulting curve fit's snuggly into the familiar
sea-shell patterns.**

Unfortunately, over time, this sacred symbol of the 3-Dimensional Phi Spiral, as most pagans know, was demonized and found it's place as 2 twisting horns on the head of the notorious Christian Devil to stain the Sacred Spiral's divinity with Fear.
Fear (**F**alse **E**vidence **A**ppearing **R**eal) means that no access to this Memory can be achieved which keeps the masses in the Dark Ages, until Consciousness begins to awaken, as is happening now.
Regarding the Demonization of the Pentacle, having placed an inverted pentagram on the Devil's Third Eye, most people understand that the divine symbol of the Witches Pentacle has also been smeared:

Fig 8
**The Pentacle, has over 200 expressions of 1:1.618
hidden within it's simple geometry.
Where any two lines intersect, the proportions happen to be
Fibonacci Numbers.
When you draw a smaller Pentacle inside the larger
Pentacle, it reduces or compresses at the rate of 1.618.
And so on forever, diminishing into the micro atom
or enlarging to the macro universe.**

The Pentacle is considered the Male Aspect of Sacred Geometry, as it consists of straight lines. Whereas, when we view yantras or power diagrams that involve curves, like Fig 2 above, it is called the Female Aspect.
The ultimate symbol of the Phi ratio is The Infamous Pentacle.
All living proteins are pent shaped.

The Dodecahedron is the 3-Dimensional form of the 2-Dimensional Pentacle, the true shape that all Pagans must resonate to is known as the 5th Element of Ether or Spirit, the most important of the 5 Platonic Solids known as the **Dodecahedron** ("do" means "2" in Greek, and "dec" means "10", as in decimal or decade, and "hedron" means face. Thus it is a 12-faced polyhedron with 12 pentagonal sides where all lines and all angles and all faces are equal and all 20 vertices touch the sphere).

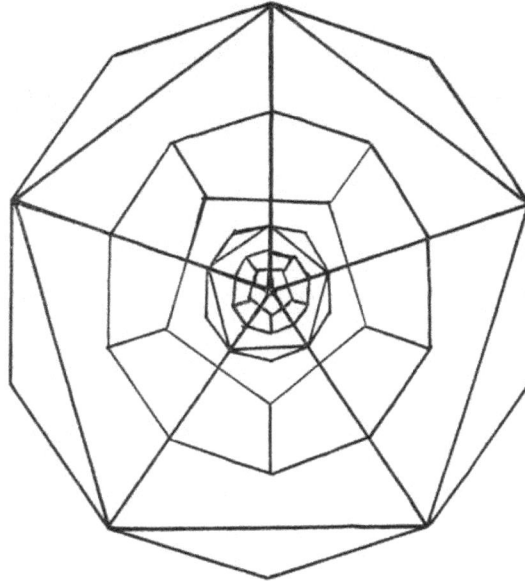

Fig. 6 — An Infinite Series

Fig 9
Infinite 1.618 Series of the Dodecahedron (12 Pentacles) and Icosahedron (20 Equiangular Triangles) nesting relationship.

The diagram Fig 9 here shows the Infinite Series of the Dodecahedron, when stellated forms a larger Icosahedron of 20 triangular sides, and when this is stellated, the icosahedron forms the dodecahedron. Thus the infinite process continues, from the atom to the galaxy, the size or shape changes, which is the key to Alchemy, but what remains the same is the wavelength of 1.618033...
This is called **Scale Invariancy**, where size does'nt matter and the proportions remain the same.
Also, the 10 sided diagram above means that the 3-D Dodecahedron has a 2-D decagonal shadow or view, in the jargon of the sacred geometrician, it is shortened to "**the dec view of the Dodeca**".

Most people know for example that the Mastercard fit's snuggly into the square of your hand, when you hold it, but if you examined the mathematics of this shape, it is a Phi Rectangle exactly 21:34.

The list of where Phi occurs in Nature, in Crystal, in Space, is endless.
After a while of studying these proportions, you begin to see it everywhere, as you walk through your town or forest. Ensure that your children learn the essence of this Living Mathematics of Nature.

Phi Art of Jain.
Harmonics of the Golden Mean
in the biometrical Human Form.
(Pencil drawing, 1982, Cedar Bay far North Queensland).

"Lost Secrets of the Phi Code"
pART 2

WHAT YOU DON'T KNOW
ABOUT PHI and THE FIBONACCI SEQUENCE:

JAIN'S DISCOVERY of The REPEATING 24 PATTERN

There is much knowledge about the Omniscient Phi Ratio, but what you don't know is also of great interest:
We have been told for thousands of years that this Phi Ratio is an infinite decimal that has no end and no visible pattern, (which is correct), that it's vibration of **1.1618033988** just keeps going on like a non-sense number, like the famous Pi of 3.141592 on and on forever without any symmetry (which is incorrect, as there does exist an invisible yet secret pattern). Conventional mathematics takes you to a certain ceiling or level of understanding, but the mathematics that is about to be revealed, is a higher level of Knowledge that will lift this ceiling of forgetfulness.
Our eyes need to be sharpened to develop a sense of X-ray eyes that are capable of distinguishing the inherent Order amidst the dominating Chaos.

But the truth is, we have been duped, or dumbed down by the conventional learning institutions. The very definition of the Golden Mean or Divine Phi Ratio is that there is a pure and marvelous symmetry embedded in the numbers, shared wavelengths that can travel from the atom to the universe without being self-destructed, in a sense that are immortal. Only the 1.618 ratio is fractal enough to ensure survival, as it knows how to be **self-similar**, **embeddable**, it knows how to be **recursive** in the micro and macro. That is, there must be a pattern visibly hidden somewhere in this infinite number. A useful image here is that of the nested Russian dolls, one within the other, from the small to the large, as a key to the DNA molecule and the spiral galaxies.
So let us say that everyone knows of the Fibonacci Sequence, but what you don't know is how this sequence relates to Time Travel and is a true Time Code.

There are 16 Vedic Mathematical Sutras, and one of them is known as "Digital Compression" which simply means we "add the digits", thus when we look at the Fibonacci Numbers: 1, 1, 2, 3, 5, 8, we notice that these are all single digits, which is our aim to express all the following digits as singularly reduced digits. Thus when we come to the next number which is 13, we add the digits: 13 = 1 + 3 = 4, and when we come to the next number which is 21, we add the digits: 21 = 2 + 1 = 3 etc. When we have effectively reduced the Fibonacci Sequence into single digits from 1 to 9, we see that it now appears like this:

1, 1, 2, 3, 5, 8, 4, 3, 7, 1, 8, 9, 8, 8, 7, 6, 4, 1, 5, 6, 2, 8, 1, 9

Fig 10
**The 24 infinitely recurring singly-reduced digits
of the Fibonacci Sequence**

This is still a meaningless sequence of 24 numbers but it begins to make sense when we examine and reduce the next 24 Fibonacci Numbers and the next set of 24. It happens quite magically that the numeric reduction of the Fibonacci series produces an infinite series of 24 repeating digits:

If you take the first 12 digits and add them to the second twelve digits and apply numeric reduction to the result, you find that they all have a value of 9.

.........PHI CODE of 12 COMPLEMENTARY PAIRS OF 9.........												
1st Set of 12 Numbers	1	1	2	3	5	8	4	3	7	1	8	9
2nd Set of 12 Numbers	8	8	7	6	4	1	5	6	2	8	1	9

Fig 11
**The 12 Complementary Pairs of 9 in the Phi Code,
Expressing the Galactic Mathematics of Base 12 and Base 9.**

These 12 pairs of 9 have a sum of **108**. (Is it mere coincidence that the external angles of the Pentacle are 108 degrees!). It means that this frequency of 108 is the hidden pulse, or rhythm that is the essence of the living mathematics of Nature, it is the reason why the Vedas worshipped this number by incorporating it into their 108 rosary beads when they chant their most famous of all mantras, The **Gayatri mantra**, the most famous Eastern song or Prayer of Enlightenment which is always chanted 108 times, also has a rhythm of 24 syllables! What else embodies mathematically these two numbers 24 and 108. It can only be the Reduced Fibonacci Sequence.

Jain's Lost Secrets of The Phi Code includes a lesson in x-raying the Fibonacci Sequence: 0, 1, 1, 2, 3, 5, 8, 13, 21, 34, 55, 89, 144 etc to discover where the infinite phi expression has it's underlying rhythm, a distinct periodicity of 24 recurring digits that can not be seen by the Western Mathematical Eye but only by the Truth of Vortex Mathematics expressed as Vedic Mathematics!

Really this Knowledge predates the Vedas and Atlantis and is better referred to as **Galactic Mathematics**.

This material shakes the whole foundations of the current Western mathematics curriculum, will rewrite our current world view on Mathematics and will require a new definition of the Golden Mean: that it's decimal does carry on infinitely but within it's infinity there is recursive or self-similar beauty. Scientists like Naudin of France and Metaphysicians like Marco Rodin wind toroidal coils in sync with this recursion or rhythm or pulse or pattern of 24 digits to get more output than their input!

Just remember that ultimately we need to ask ourselves, what shape encodes this quality of 24, what shape has 24 faces and 24 edges. There is only one shape that does this, and it is the atomic structure of Silicon Chip which is a Star Tetrahedron, or two interpenetrating or inter-digitating Tetrahedrons, that forms the basis of many crystals. When you join the 8 vertices of this shape, you get the Cube.

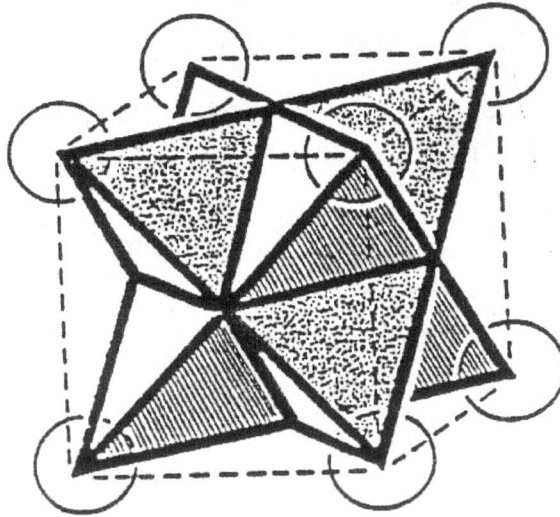

Fig 12

**The Star Tetrahedron, which is the 3-Dimensional form of
the Star of David, is the geometric equivalent to the Phi
Code expressing precisely the 24ness exhibited
in the Reduction or Compression of the Fibonacci Numbers.**

Basically, **Shape stores Memory**. Due to this profound
mathematical discovery, there is now established a mathematical
and spiritual connection between Technology and Consciousness,
between Sacred Geometry and Inter-Dimensional Time Travel. This
Number 24, this StarGate, this clock of 24 Hours, is what Einstein
called the Fourth Dimension: Time.

If this number "24" cropped up again in another Mathematical
Sequence, would you be convinced that this is no mere co-
incidence.
According to a German chemist, Dr Peter Plichta, the Egyptians
knew the secret to the hidden pattern of Prime Numbers. Imagine if
you wrote the numbers from 1 to 24 in a circle, going
clockwise, and the next 24 numbers concentrically around it, and
you repeated these many rings of 24 consecutive numbers, have a
guess where all the prime numbers would lie?
You guessed it, on 4 distinct diagonals or diameters that form the
4th Dimensional Templar Cross.

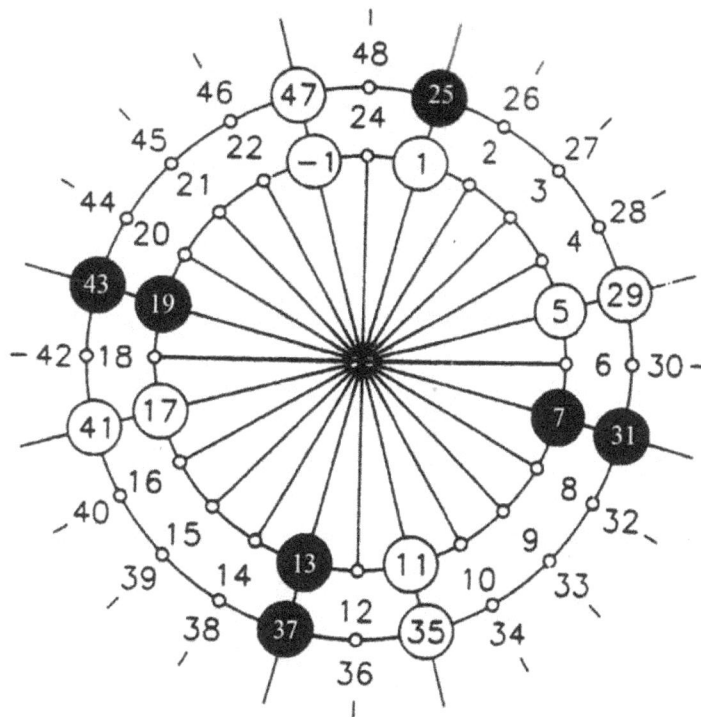

Fig 13
**The Symmetry of the Prime Numbers
shows the 4th Dimensional Templar Cross
based upon concentric rings of 24 consecutive numbers.**

Yet again, all the academic books are embarrassingly wrong, they educate our children that the Prime Number Series has no meaning nor pattern. I wonder why our military forces uses the higher end spectrum of this Prime Number Series for advanced encryption technology.

For those readers who are aware of the Philadelphia Experiments that aimed to turn military naval boats invisible during times of war, did you know that the science behind it was purely working with the Fibonacci Numbers.

Did you know that the total success of the modern computer age is dedicated to the Fibonacci Numbers, for the ability to send large files, from one system to another, anywhere in the world, depends on the ability to compress files? (which is what the sunflower does with it's seeds, they compress into the 21:34 ratio, otherwise, if the seeds along the counter-rotating spiral arms were arranged say equally at 21:21, they would simply fall out of their floret). Thus computer designers, by emulating Nature's Pure Principle, solved their problems thanks to what we call Fractal Compression, which as a frequency of numbers is the pure 1.618033988 consciousness. Remember that Phi is not a Number, but rather a "cascading of frequencies" of the Fibonacci Numbers.

Yet why are we not educating our children about this fountainhead of Knowledge?

I have been commissioned all over Australia to teach children, over the last 20 years, and globally over the last 2 years, with one simple job: to teach the Beauty of Mathematics.

It's quite an evolution of new mathematical gems coming through, and if you stay in tune with this website, **www.jainmathemagics.com** I will have more of these essential and psycho-active diagrams posted for you. These articles will help you discover that Maths is Art, Maths is Science and Maths is History. They will help you to explore Sacred Geometries that are invisibly constructed and nested within the Heart, and ultimately teaching you that all Knowledge is within You. You only need to learn how to **Remember** this Knowledge, it is already there.

The most important element in Sacred Geometry is that we already know this information, that the process is about remembering this Lost Knowledge of the PHI CODE that is already morphologically and geometrically encoded in your bio-magnetic and bio-electrical fields.

Essentially, the human body is in resonance with the **Living Mathematics of Nature.** Knowing this, is part of your Ascension Process.

Ultimately, this Pythagorean Knowledge is not an intellectual process, but rather that these essential geometries are invisibly constructed and nested within the Heart as the Centre of a greater Galactic Grid connecting us to all Memory of all Universes and Atomic Levels. Learn to sit still everyday, and in your daily meditations there is one true key to mastery, and that is to give thanks for all that we already have: give Gratitude.

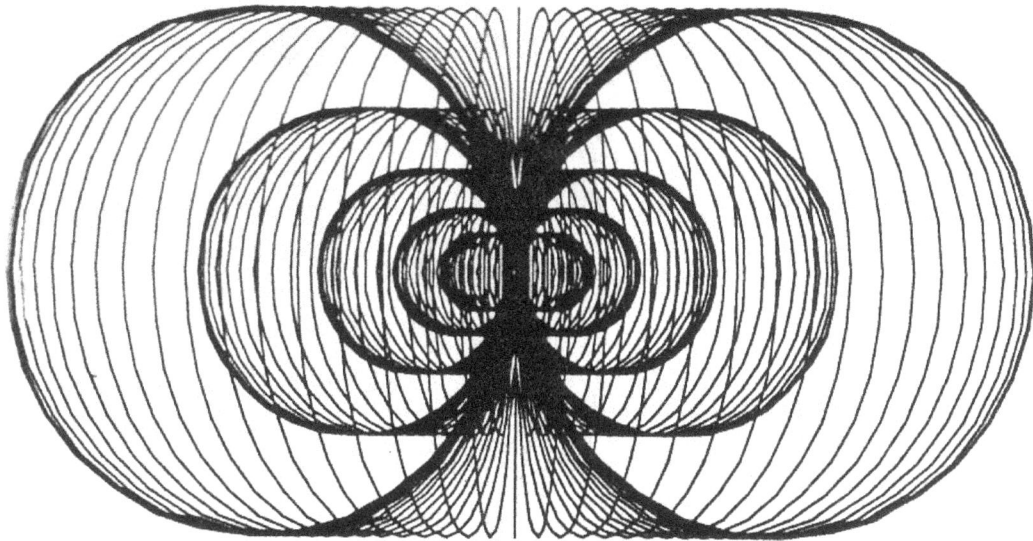

Fig 14

Transverse cut of the Torus, showing the nesting of many Tori, and separated by distances of the Fibonacci Numbers. The Wormhole Centre that bridges the Worlds is your Heart. This material is classified as the Advanced Knowledge on the 12:24:60 Phi Code, the Knowledge of Sharing.

In the next article/lecture, we will learn the advanced and practical use of integrating this ancient Knowledge of the Infinitely Repeating 24 Phi Code by breathing in the Memory of the Celestial (that which is above), and the Terrestrial (that which is below) into our Heart. The meeting place of these counter-rotating fields, like the dynamics of the Sunflower floret witnessed in Nature, is One Breath meeting in the Heart. A One Breath Heart Meditation, copying and imitating Nature, is the fulfillment of this intellectual knowledge so that it becomes an Experience, an unlimited Journey through Time and Space, navigating with the powerful tools of Numbers, Fixed Eternal Design. This Meditation, the key subject that I teach around the world, and now to Children, is called the **EARTH-HEART Meditation**.
It's the key to the Physics of Time Bending (Travel).

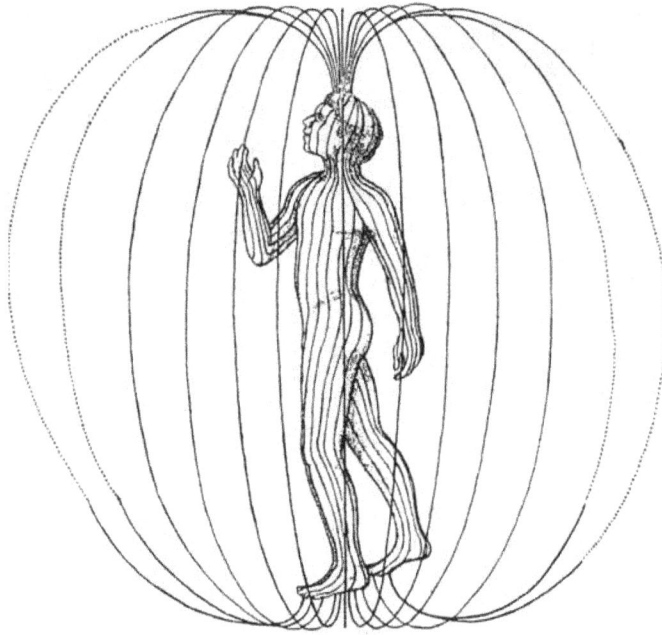

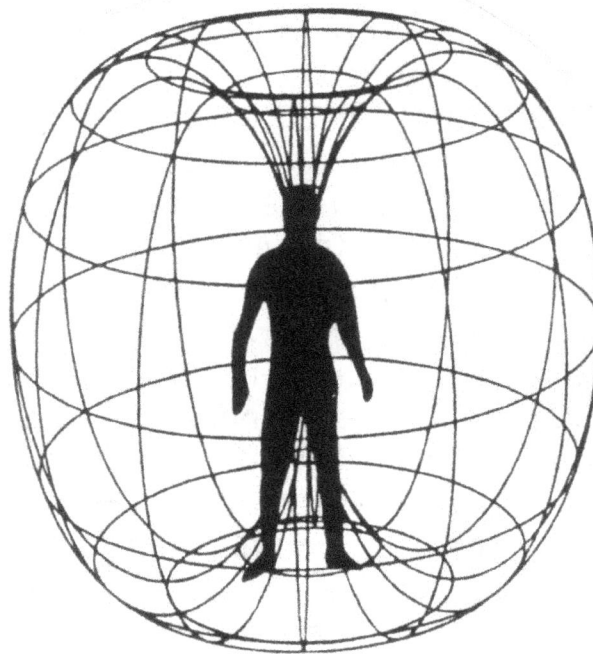

Fig 15
Jain's Earth-Heart Meditations
is the embodiment of this Ancient Knowledge,
by simple, effortless in-breaths into the Heart, and out-
breaths to your surrounding domain, expanding to the
spherical diameters of the Fibonacci Sequence.

pART 3

WHAT YOU NEED TO KNOW
ABOUT PHI and The FIBONACCI SEQUENCE:

TIME CODE 12:24:60 ENCRYPTED
in the FIBONACCI SEQUENCE
and PASCAL'S TRIANGLE

Having established that all the books declaring no mathematical phi recursiveness are wrong and that there is indeed immense symmetry in the Fibonacci Sequence, specifically an infinite repeating pattern of 24 reduced digits, it is in this section that we stumble upon another important cycle of repeating digits. Before I give it to you, you must ask yourself:

"What number or numbers, symbolically, represent 'TIME' to you?"

Many people will say, 12 or 24 hours or 60 seconds to the minute or 60 minutes to the hour, which is the vestigial remains of the Babylonian use of Base 60 several thousands of years ago etc

What if, in this article, I could link these critical numbers:

12 : 24 : 60 in a manner that connects them to this concept of Time Codes to Nature. Would you be impressed?

In Part 2, I showed how the number 24, symbol of Hours in the Day, is hidden in the Fibonacci Sequence, thus this section of Part 3 will now focus on where the Cycles of 60 reside.

"The Cycle of 60 Pattern" in the Fibonacci Sequence:
I am rewriting the same sequence again, but this time I will highlight only the **final digits** of the Fibonacci Sequence:

0	1	1	2	3	5	8	13	21	34	55	89	144	233	377	610	987	1597
Fig 16																		
The Final Digits of the Fibonacci Sequence, highlighted																		

This time, I will only write out the final digits:

0	1	1	2	3	5	8	3	1	4	5	9	4	3	7	0	7	7

Fig 17

**The final digits of the Fibonacci Sequence
display a Periodicity or Cycle Length of 60**

There is no point asking you to observe if you can see a pattern as it turns out that the series is **60** single digits long, and that these **60 digits keep repeating forever** and forever in that same series. In mathematics, we say that the series of final digits has a **periodicity of 60** or that it has a distinct cycle length of 60.
It means, that this cycle repeats itself after Fib(60).
So here it is, another hidden, infinite yet simple code based on 60.
So next time someone asks you: "What is the Time?" you can present to them the pattern in this series.

Here is a diagram Fig 17a that I have hand drawn, to show you these 60 last digits written into a wheel. Since it is based on 60, I have also written an outer layer of writing that says EARTH EARTH written 12 times, which creates my logo: EARTHHEART, the purpose being to program or imbue this EARTHHEART with the vibration the deep structure coding of the Golden Mean Mysteries.

Just remember that this Wheel of 60 is Fixed Eternal Design, or call it Timeless Design, that our concept of Time, the 4th Dimension, was structured on 60ness and also upon the Laws of Nature since the Fibonacci Sequence is a direct link to how Nature invisibly operates.

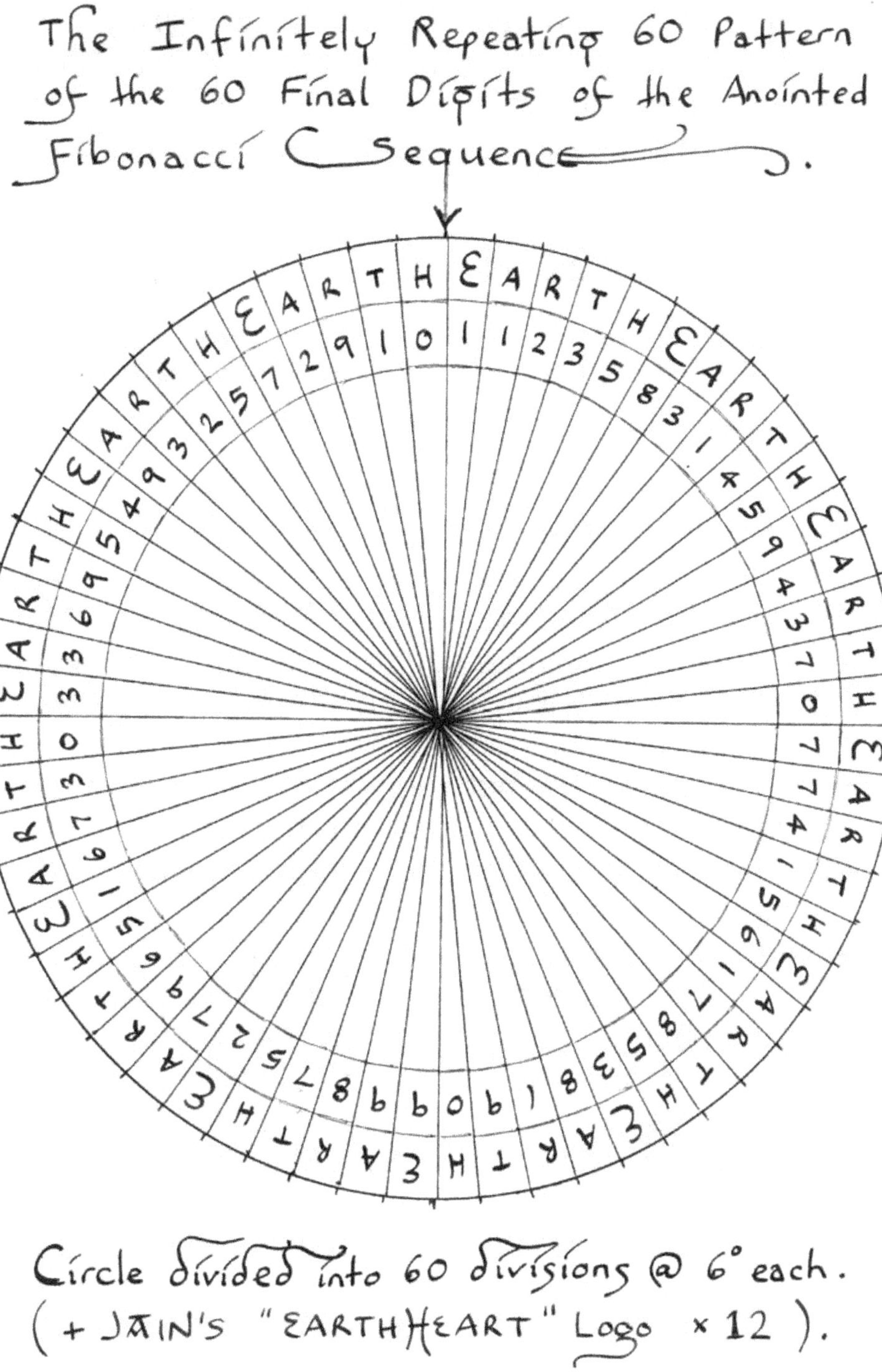

The Infinitely Repeating 60 Pattern
of the 60 Final Digits of the Anointed
Fibonacci (Sequence).

Circle divided into 60 divisions @ 6° each.
(+ JAIN's "EARTHHEART" Logo × 12).

Fig 17a
The Wheel of 60
showing the 60 Final Digits of the infinite Fibonacci
Sequence that has what we call a distinct Period or
Recursion or Periodicity of 60.

But it gets even better.

What if we were to examine all the **final two digits** in the Fibonacci Sequence and inspect for another pattern. If there is one, what would it's periodicity be?

That means I need to rewrite the sequence and highlight the final two digits:

0	1	1	2	3	5	8	13	21	34	55	89	144	233	377	610	987	1597
Fig 18																		
The final two digits in the Fibonacci Sequence has a Periodicity of 300																		

Remarkably, there is yet another distinct recursion happening, but since it is too long to write out, I will just state that it happens every 300 digits long. That means after Fib(300), the final two digits keep repeating the same sequence again, then again then again, forever. The periodicity or cycle length is 300.

(Before I give you more hidden patterns, at this stage, I would like you to note the relationship between the two cycles of 60 and 300. We can summarize that as a proportion, they are in a 1:5 ratio. This One to Fiveness proportion has a lot to do with the Pentacle Harmonics and the Human design, 5 fingers on each hand and 5 toes on each foot).

Let us continue. The fact that we have discovered two incredible patterns in the final digits and then with the two final digits, you could easily surmise now that there may be a good chance that there exists patterns in:

the **last three digits** and the **last four digits**, and the **last five digits**.

Again, the Patterns are clearly there, thanks to the help of modern computers. Actually, this article is therefore a tribute to the fantastic world of modern computers our Tech God, that reveals the infinite sublimeness of pattern recursion.

It can be realized that there are more and more patterns:

- For the last **three** digits, the Periodicity or Cycle Length is **1,500**
- for the last **four** digits, the Periodicity or Cycle Length is **15,000** and
- for the last **five** digits the Periodicity or Cycle Length is **150,000**

and so on...

Since we are avid Pattern Hunters, we want more patterns and links, as now it is becoming quite a serious matter, that Nature is not random, but obviously there is some inherent Order in how the Universe operates. One thing that cannot lie, is Proportion, as it is a forever science or a universal language.
Let us therefore list these important numbers, and examine their proportions to one another:
60 : 300 : 1,500 : 15,000 : 150,000
We can now simplify these ratioes, by dividing every number by 60:
1 : 5 : 25 : 250 : 2500
which appears to increase by fives then tens powers.

Let us now compare each number to the one adjacent to it:
60:300 = 1:5
300:1,500 = 1:5
1,500: 15,000 = 1:10
15,000:150,000 = 1:10

To continue this investigation, in our attempt to link these ratios to other important mathematical sequences, we will list the above ratios in this format:
1 : 5 : 10 : 10

Obviously, we could explore this sequence further, but again we need the power of the computer.
Part of the Jain Mathemagics Institute or Temple of Mathematics, is to locate these out of reach codings, and if there be any truth in the importance and grandeur of Phi, then surely we can as a community of interested souls, academics, professors, mathematicians, nerds, geeks, hackers, all sorts of highly

intelligent folk, would be tempted to determine the furthest outreaches of symmetry lurking in the depths of our creation story.

And what if the fundamental basis for this underlying or hidden mathematical structure was known by the ancient Vedic scholars. If the Indian scholars from 2,000 years ago knew of this Fibonacci Sequence, would we by right acknowledgement be forced to change the name to the original discoverers of the Fibonacci Code. According to Vedic texts, the Fibonacci Sequence was attributed to HemaChandra-Gopal well before it arrived to Europe via Arab wars and traders. Fibonacci was living in the time of the 12th Century, Europe, and HemaChandra-Gopal lived about 2,000 years ago in Bharat or India. Do we now call the Fibonacci Sequence the HemaChandra-Gopal Code. Or do we just accept that there were other pre-Vedic cultures, and that all Knowledge has existed, and no-one can ever really put their name to any discovery which is only a Re-Discovery. Thinking along these lines, there is another famous pattern called Pascal's Triangle, but it also has it's so-called origins in ancient India.
The famous Meru Prastera Pattern is also mistakenly called Pascal's Triangle. (The image of "Meru" is like a holy mountain, like Mt Olympus is to the Greek Gods).

						1						
					1		1					
				1		2		1				
			1		3		3		1			
		1		4		6		4		1		
	1		**5**		**10**		**10**		**5**		**1**	
1		6		15		20		15		6		1
					Fig 19							
			The famous Meru Prastera Pattern or Code aka Pascal's Triangle									

Before explaining what Pascal's Triangle is all about, and how to create it, let me immediately connect the previous code of **1 : 5 : 10 : 10** to be found in the above Pascal's Triangle, along the 6th horizontal line. This discovery basically connects the Living Mathematics of Nature, which is the Fibonacci Sequence, to other mathematical giants of Inherent Order and Underlying Symmetry like Pascal's Triangle.

All is connected. All is a Unified Field. All is One.

These two important sequences establishes a direct connection to both the Fibonacci Sequence aka HemaChandra-Gopal Sequence and the Meru Prastera Triangle, revealing that they are intimately connected to the **12 : 24 : 60** Galactic Time Code.

35. Detail from a Koran frontispiece with a design laid out on a grid similar, though not identical, to that shown OPPOSITE. Egypt, 1356.

Islamic Dodecagram (12 pointed star) motif found in the Egypt, the Koran, circa 1356AD

TIME CODES 12:24:60
Or
THE TWENTY-FOUR GODS THAT ARE ONE

I ask You, the audience:

What numbers, symbolically, represent "Time" to you?

Please call out any numbers that you know to be relevant.

Many people will say, 12 or 24 hours or 60 secs to the minute etc

What if, in this lecture that I could link up to you, these critical numbers:

!2 : 24 : 60

in a manner that connects them to this concept of The Underlying Time Code to Nature, (and with the Physics of Time Bending) with Nature being symbolized by the Divine Proportion 1:1.618...

Would you be impressed?

And what if the fundamental basis for this underlying or hidden mathematical structure was known by the ancients Veda scholars and represented as the HemaChandra-Gopal Code that we now mistakenly call the Fibonacci Sequence!

The famous HemaChandra-Gopal Pattern is also mistakenly called Pascal's Triangle:

By definition, how to create Pascal's Triangle:

Each entry in the triangle is the sum of the two numbers in the row above. A blank space can be taken as "0" so that each row starts and ends with "1".

Thus you can see that "1 + 1" = "2" and

"1 + 3" = 4 etc

In fact, most PhiBonatics already know that the diagonals in Pascal's Triangle have a sum relating to the Fibonacci Numbers.

The Fibonacci Numbers in Pascal's Triangle

```
1
1   1
1   2   1
1   3   3   1
1   4   6   4   1
1   5   10  10  5   1
1   6   15  20  15  6   1
```

Fig 20
Re-arrangement of the Meru Prastera Triangle
All rows are now Left-Aligned.
Notice how the Diagonals have sums which are the Fibonacci numbers?

To show you this clearly, this time I am going to highlight three diagonals in Bold, and 2 diagonals underlined, so that you can begin to see the first pattern:

```
1
1   1
1   2   1
1   3   3   1
1   4   6   4   1
1   5   10  10  5   1
1   6   15  20  15  6   1
```

Fig 21
The Diagonals of the Pascal's Triangle
create the Fibonacci Numbers.

Highlighting the Diagonals to reveal the Fibonacci Sums
Can you see in bold & underlined that 1 + 3 + 1 = **5**
Can you see in bold only that 1 + 4 + 3 = **8**
Can you see in bold & underlined that 1 + 5 + 6 + 1 = **13**

The numbers of the diagonals continue like this forever! Always adding to a Fibonacci Number, in order.

In mathematics, it is important to submit these findings as a Mathematical Paper, so others can share it. Thus, the general principle that we have just illustrated here is:

"The sum of the numbers on one diagonal is the sum of the numbers on the previous two diagonals."

Another arrangement of Pascal's Triangle:

Perhaps a clearer way of demonstrating the same principle is to have it represented more linearly, having the numbers simply tabulated one under the other, but with one change, by sliding all the rows over by 1 place:

1	2	3	4	5	6	7	8	9	10
1									
	1	1							
		1	2	1					
			1	3	3	1			
				1	4	6	4	1	
					1	5	10	10	5
						1	6	15	20
							1	7	21
								1	8
									1
1	1	2	3	5	8	13	21	34	55

Fig 22

Alternative diagram illustrating how the Diagonals of the Pascal's Triangle create the Fibonacci Numbers. The Bottom Row in bold is the sum of the digits posited in the vertical columns.

OTHER INTERESTING FACTS
ABOUT PASCAL'S TRIANGLE

Meru Prastera Triangle has lots of uses including:

1 - To solve problems in Probability.
Imagine you were asked how many permutations of Heads (H) and Tails (T) or different ways are there of throwing 3 coins onto a surface, we know that the answer is:
2 to the power of 3 which is $2^3 = 8$.
But if you were specifically asked what are your chances of getting a Head (H) in any throw how would you record your entries to fully analyze the data?

This would be the solution:

3 heads: H=3 is found in **1** way (H H H)
2 heads: H=2 can be got in **3** ways (H H T, H T H and T H H)
1 head: H=1 is also found in **3** possible ways
(H T T, T H T, T T H)
0 heads: H=0 (i.e.: all Tails) is also possible in just **1** way:
(T T T)

Do you notice the highlighted numbers in bold: **1 3   3   1**.

It is the 3rd horizontal row of the Meru Prastera Triangle.

2 – The Meru Prastera Triangle can also help find terms in a Binomial expansion: (a+b)^n
eg: $(a+b)^3 = \mathbf{1}a^3 + \mathbf{3}a^2b + \mathbf{3}ab^2 + \mathbf{1}b^3$

I will not go into the theory here. I have merely set the ground work to introduce you to the amazing worlds of the HemaChandra-Gopal / Fibonacci Sequence and the Meru Prastera / Pascal's Triangle.

CONCLUSION

What now arises for me as an adamant researcher, is the contradiction going on between this important and timeless, universal discovery, of **12:24:60** and how it conflicts with the much flaunted and bantered new time code called the Mayan Calendar whose main proposition is **13:20**.

Although the current twisted literature on the Mayan Calendar is exceptionally popular, it must be reviewed and reconsidered. It is written by Jose Arguelles, an author, artist and visionary. Their cult's key catch phrase is: 12:60 = TIME IS MONEY (the old paradigm that we are collectively shifting out of) and 13:20 = TIME IS ART (the new paradigm that we are realigning ourselves to). I have met some of the Mayan Elders and they are very upset by this westernized distortion that fabricates "days out of time" and other sensational claims. I am concerned also that galactic Base 12 is now under severe "bad press" and am warning young sacred geometricians to be careful of this Mayan trap. In my hometowns of Byron Bay and sweet Mullumbimby many people are worshipping this 13 full moon per year invention and blindly debunk anything to do that references 12 or 60; they therefore can not see the divine connections to 12 and 60 in the phi mysteries.

Arguelle's premise is strongly challenged by <u>Dr Bernard Jenkins</u>, (and myself), illuminating all his readers that Jose's time count is wrong, weird and fudged, and that there are not 13 full-moons in the year.

Actually, we need to know exactly how many full moons there are in the year. I have researched that there are not 12 nor 13, but somewhere in the middle, like 12.4. An astronomer would agree with this. The purpose of this article, is to not believe all the literature you read. We are Pattern Hunters, looking and investigating for clues that are based on timeless laws of the Universe's cycles. Yes we know that the Fibonacci Sequence is fully laden with the cycles or periodicities of 12:24:60 and that is definitely connected to the Meru Prastera Triangle, and thus is a clue to Time Travel, Space Travel, Inter-Dimensional Travel... that is the next step, linking this Divine Mathematics to the Divine Sciences.

So I live in an area in Byron Bay which has the potential to be a new model rural and coastal town that is aiming to be self-sustainable for the future in terms of recycling our own waste, buildings with no chemicals, cleansing the water-ways, and introducing this Vedic and JainMathemagics into our schools. But all my peers are waving the Mayan Calendar flag subtly proclaiming that Base 12 is bad news, and hey man, we got to shift to the 13 frequency etc but how can I accept this bad press **when I am armed with this irrefutable evidence that Base 12:24:60 is It, was It, and will always be It**.

One of the important warnings of our life-time, I believe, is to not get caught up in the 2012 prophesies, it will become a madness in the next few years, and create so much Fear that it will globally dis-empower our Forces to bring real and effective change into this world. Since 2012 is future-based, it will only serve to keep us out of The Now, which is our real power, The Now. And Fear will be generated in every nation, that ends times are coming. Just like what happened at the turn of the bi-millennium, panic everywhere, people selling houses, Y2K bug created false fear etc. So bewarned, stay in your Heart and **if anyone raises the 2012 flag or banner just walk away**, do not even engage in a conversation with them as they are over-inflamed and superficially correct. The wise person stays in The Now, and returns to the Garden where the true Mathematics is stored.

No doubt, the prophesies will come true, that is the changes, they will happen, but the discussion here is that Man-Woman does not know how to count Time, it has been infected with a virus, wars and many cultures have interfered with the original blueprints, so 2012 will happen, but when you are looking the other way. And all it will ever mean is a raising of consciousness at the expense of racial cleansing. Over-population must be reduced, climates must rebalance, the waters must be renewed from chemicals etc so the times they are a changing, but no calendar can predict it.

There can be no argument about it. If there were precisely 13 full moons in our year, then I would be in a position to reconsider my findings, but I can not budge.

We must keep doing the research. Go to any astronomer and enquire, how many full moons are there in the solar year, sidereal year, and how do you measure this etc... Ask other researchers and compare their findings.

The other fascinating evidence is from the book called:
"**CIVILIZATION ONE**" by Christopher Knight and Alan Butler (www.civilizationone.com) which states quite clearly that our forbears did not compute their calendars and weights and measures based on the solar year of 365 days but rather chose to index their data against the stars, and Venus' orbit, and concluded that we actually have always used **366** days per Sidereal Year (sidereal meaning the stars). Now this is a revolution in Mathematics and Astronomy. It challenges everything you ever believed about calendrical time count. Any person or book that can shake your current belief systems is to be welcomed. We must remain open to these occasional waves of new knowledge that surface above the other waves in this ocean of infinite belief systems. But one system, above all others is a universal wave, an omniscient wave that is always there, but hidden in the www. of infinite possibilities, it is the science of Harmonics or Cycles or Proportions. Simply: Mathematics. When Pythagoras's works were retranslated, there arose a debate about the translation of the Greek word: "Logos". Many of you are familiar with the Christian expression: "In the Beginning was God" but it has been interpreted as "**In the Beginning was the Word, was the Sound, was the Vibration**" etc many meanings, as to what the actual word "Logos" meant. According to Greek scholars,
Logos = Proportion or Ratio. Mathematical Proportion, as in our case or enquiry of **12:24:60**.

May this article be of some inspiration to You, at least to always weigh the evidence and to give priority to ancient Mathematical Revelation which can not lie.
Mathematics, or Logos / Proportion is the common language for the new globalizing and **fibonnacization** of the world we live in. It is time to restore it. And Mathematics is the supreme language, not French or German which, no offence, are decaying rapidly, they are limited. As much as they are precious cultures, like all cultures, we can not, on the other side of the world relate to them or understand them. But what is in common, now, for the last 10,000 years, and for the next 10,000 years to come is Mathematics. We will still have this Mathematics of **12:24:60** on different planets, on Mars and Neptune, how can it ever change, this **FIXED DESIGN**, this **ETERNAL PROPORTION**.

May all our children learn this Jain Mathemagics and become acquainted in the class room with these precious jewels known as:

- ➤ The Fibonacci Sequence aka The HemaChandra-Gopal Sequence
- ➤ The Pascal's Triangle aka Meru Prastera Triangle
- ➤ The Prime Numbers 4th-Dimensional Cross Symmetry of 24 Circular Numbers
- ➤ The Squaring of the Circle aka the Mystical Quadrature of the Circle
- ➤ The Vesica Piscis
- ➤ The True Value of Pi which is a revolution in Mathematics (Jain Pi = 3.144... based on the Square Root of Phi)
- ➤ The 24 lengths and 24 faces of the Star Tetrahedron
- ➤ The Value of Intuitive Mathematics

Finally, I would like to discourse on the symbolic meaning of the beginning of the Fibonacci Sequence:
0, 1, 1 and the Journey to Infinity:
What does this repetition of the "**1, 1**" truly mean.
We know that "0" is the Void, Source, Bindu, Emptiness, yet The All. It is the crucible or womb for the possibility of any God Manifestation.
If "1" represents "God", then the occurrence of another "1" suggests that In The Beginning, "God" duplicated Him-Her-Self, to see reflected the Divinity of Self. God-Dess-Ence needed to make Him-Her-Self **SELF-SIMILAR**, thus the "1" became the "1" (not the traditional view that the "1" became the "2" which meant from Oneness we entered the Die-Mention of Duality, no, there was never ever any real separation as we are in all things and the true meaning of ascension IS TO SEE GOD IN ALL). Thus God-Dess-Essence embedded Her-His **Unity-Consciousness** into all things and journeyed or quested in search of the ultimate light vehicle possible for great expansion or contraction/compression which was the Sphere in the Process of Implosion or the harnessing of the **Toroidal Domain**. (This only means that the shape of your blood platelets are doughnut-shaped!). God-Dess-Essence needed to keep multiplying without being destroyed, so chose the simplest of all arithmetical (adding) and geometrical (multiplying) sequences, combined them and thus the infinite process forever continued when 1+1 became 2 and the 1+2 became 3 and 2+3 became 5, and the 3 + 5 became 8, a mathematical wavelength concept

known as **SHARING**. (This only means the special place where the elbow bends, or where the knee bends, or the mathematics of the belly-button). So well disguised was this Unity Consciousness of "**1,1**" that it could **EMBED** in the atom or the stars, in the floret of the sunflower or the pinecone, but ultimately it's secret head-quarter's office is the spiral staircase that is your **D.N.A** that allows access to the micro and macro worlds by sensitive understanding of The Laws of Time and Space that we call Proportion or Logos and is mathematically described as 12:24:60, a Trinity of Frequencies that obeys the embrace of Mother Earth and Father Sky who teach their Children of the Crystal, Mineral, Star and Earth Domains, to not only learn the Physics of Time Bending and Astral Hygiene, but to learn how to fractally embed this Knowledge in their Heart. (Though this is only a belief. How do we integrate the fact that our native Australian Aboriginal Elders of the Dreamtime declare that the Sun, is Female!).

How do we make the essential **relationShift** from the single Head of strict belief, to the open source of Twin Hearts.

So really, as the logical left-brained men of this world analyze and rationalize and write such articles as this, a right-brained intuitive woman could merely smell the Rose with such genuine depth and feeling that this mere act of smelling could access all Knowledge. "How beautiful the nesting and embedding of those petals in such gorgeous symmetry and chaos", she surmises. Thus the time of evolution has brought Man and Woman to become "1 and 1" or Androgynous, time for Men to learn how to Smell the Rose, and learn how to Cry, and Woman to lay the bricks and master Mathematics mentally, such that there is no distinction between the faculties of Man and Woman.

This article I pray will help heal any agonizing memories of being mathematically wounded. If you disliked mathematics as a school student, you probably did the right thing by avoiding it, as your Higher Self switched off in fear of being bombarded by incorrect conditioning, so that is good. If you were deemed "**dyslexic**" at school, then consider yourself Gifted. Our well-intentioned teachers failed to address us visually and bombarded us with meaningless simultaneous equations and differential calculus that had no meaning for us, so we smartly switched off in the classroom, and for mere neural entertainment we began to flip, rotate, reverse and mirror-image the tirade of numerical data for sanity's sake. So congratulations if you were dubbed dyslexic or dumb, as your Higher Self's thermostat had to switch off to avoid the excessive conditioning. But it is time now to restore the imbalance and recognize that all Mathematics is a Sacred and Universal Star Language. It only needs now to be taught coherently, simply and visually in all schools of the world so that it switches on the remaining dormant codons of DNA igniting a child's consciousness to remember that once Mathematics was considered **a most Beautiful and Anointed subject**.

Thus in summary, we can conclude that our ancient cultures had a great understanding of the harmonics that pervade the universe. Typically, the two most important mathematical bases are Base 9 and Base 12. (12 x 9 = 108).
The one and only cosmology that obviously depicts both these bases is the Sino-Tibetan calendar:

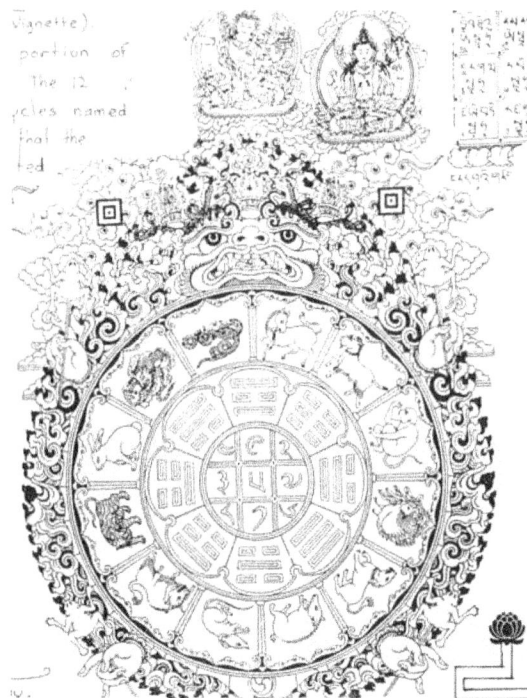

Fig 23
The Sino-Tibetan Cosmology or Calendar depicting the outer ring of 12 animal signs girdling the central motif of the Magic Square of 3x3 composed of 9 cells or the first 9 counting numbers.

So why was this calendar so important, showing a great god: **Manjushri**, incarnating as a fiery tortoise whose energy is to protect the sacred center. And why is the Lo-Shu or Magic Square of 3x3 in the center of the cosmogram, why is it not just a small intellectual motif in a corner somewhere? I believe that the combination of both the **9 numbers of the Lo-Shu** and the ring of **12 animal signs**, act like a sophisticated padlock or combination key ringing to the sound of 9 x 12 which equals this **magic number 108** so far discussed. Does such a calendar place importance upon Time, in the sense that it refers to a specific angle in the sky when and where you were born astrologically, and if so, is this eternal recursion of the 108 harmonic of the Phi Code really a **Stargate**? Is this Phi Code really a Time Portal? Many questions abound from these simple observations as found in Nature and Mathematics. Why would a whole culture, known as the Vedas of the ancient Indian tradition worship the number 108 and not even know why they are doing so? How many thousands of years has this secret been whispered? And without these mathematical sutras or links, there appears to be something **missing** in our understanding of the whol-:

Th- conclusion that I am making h-r- is that a lot of th-s- qu-stions could not b- ask-d had it not b--n for th- ability to und-rstand th- most simpl- V-dic Sutra call-d Digital Sums or Digital Compr-ssion, bas-d on th- simple arithm-tical fact that 34 = 3 + 4 = 7, this x-raying of imm-ns- data b-ing r-duc-d to fundam-ntal harmonics is th- subj-ct of d--p appr-ciation h-r-, that our childr-n n--d to r-m-mb-r that Math-matics is a supr-m- and Univ-rsal Languag- and it can be dir-ctly acc-ss-d by sm-lling th- flor-t of a ros- or a sunflow-r.

R-gards **JAIN**

As a follow up to this series of articles,
Jain will be giving a one day seminar
that will illuminate this body of knowledge.
entitled:
"The Vedic Mystery of 108 Revealed"
on the 1st October, 2006, in Byron Bay's Abraxas Bookshop.

for more information on this weekend seminar, and others,
please click on:
http://www.jainmathemagics.com/category/5/default.asp
Since his return from India, Jain has been hailed as the first
western scholar to explain to the eastern world the deeper meaning
of the **holy sri 108**.
The Indians are now calling him by the name of Jain 108.

Testimonial:
"*Sunday was so exciting, the revelation you receive for the
recursive Phi pattern is at once so staggering... yet excitingly
familiar, my physical perception actually shifted. The potential of
your rediscovery leaves me breathless. Thank you for bringing the
joy of this pure science into our lives*".

Ingrid Burke, Sunshine Coast, Australia.

Jain 108
Behind him is a typical cloth banner
of a Magic Square of 4x4
rotated upon itself at 4x45 degrees.

Email Contact: **jain@jainmathemagics.com**
website: www.jainmathemagics.com
        ✛ www.mathemagicsasia.com

For more detailed information on the Lost Secrets of the Phi Code,
You can order from Jain's website, two highly informative books
and 3 dvds included in a 5 dvd set:

THE BOOK OF PHI, Volume 1
Subtitled: The Living Mathematics of Nature.
$60   172 pages

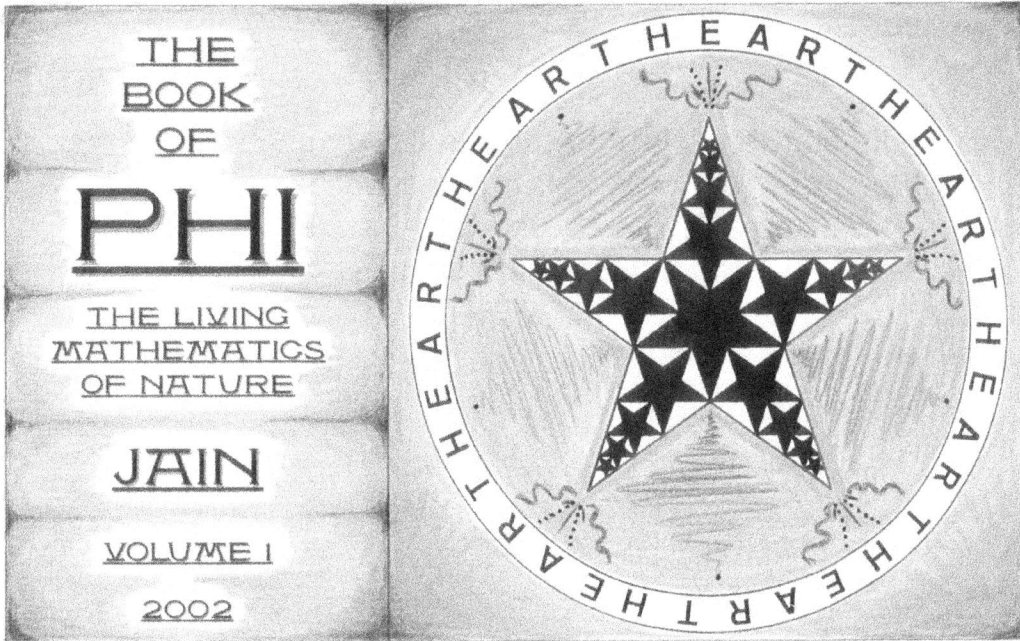

THE BOOK OF PHI, Volume 2
Subtitled: In The Next Dimension.
$60   188 pages.

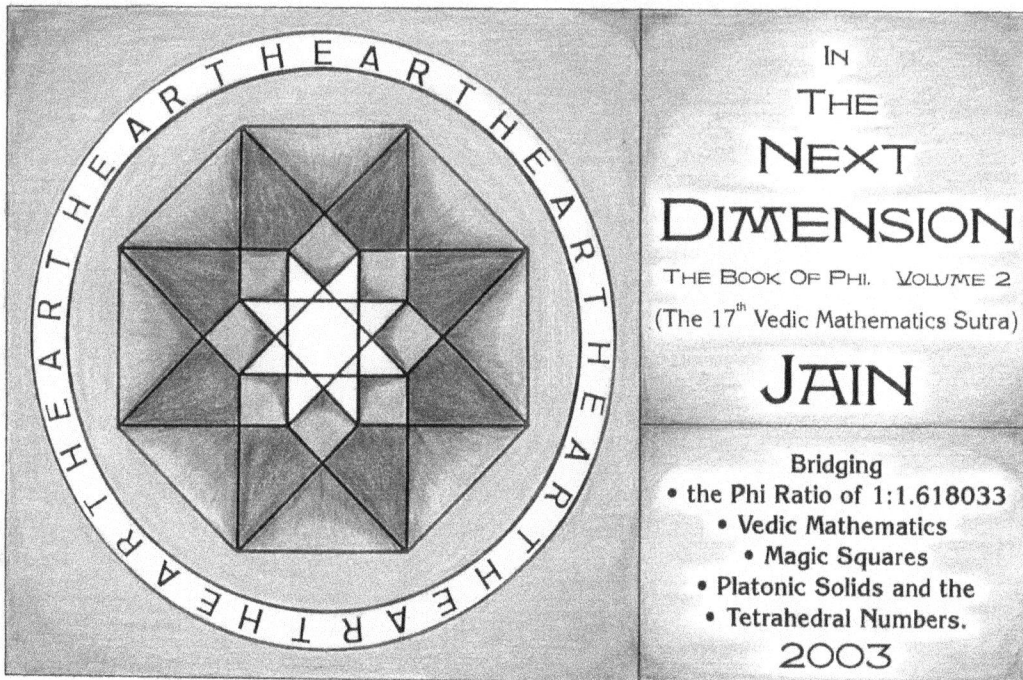

5 DVD Set – THE LIVING MATHEMATICS OF NATURE, Series
2 Hours each dvd. $200 for 10 hours of my life's work
Part 1 of 5: Introduction. Excellent visuals and explanations.This is the only dvd that gives a total overview of this Mathemagical body of work and is a great introduction explaining sacred geometry.
Part 2 of 5: Vedic Mathematics: Rapid mental Calculation
Part 3 of 5: The Atomic Art of Magic Squares
Part 4 of 5: The Divine Phi Proportion
Part 5 of 5: 3-Dimensional Sacred Geometry: A detailed description of the 5 Platonic Solids and some of the 13 Archimedean Solids.

Distributed by Jain Mathemagics
The **Living Mathematics Of Nature**
5 DISC DVD SERIES
New DVD Series

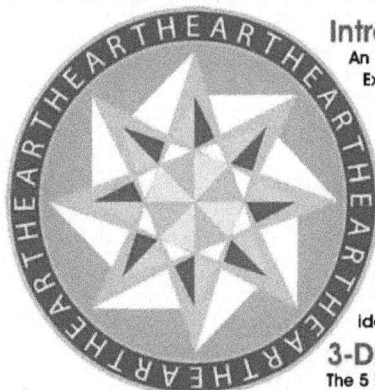

Introduction to Jain Mathemagics
An introduction to the 4 topics below.
Explains each topic, providing good overview of work.

Vedic Mathematics
Rapid Mental Calculation.
No more calculators. Increases your memory power!

Magic Squares
Translating numbers into Atomic Art.
Teaches children pattern recognition.

The Divine Phi Proportion
Explores the Geometry of Flowers identical to the Human Canon!

3-Dimensional Geometry
The 5 Platonic Solids adored by Pythagoras and his community.

Distributed by JAIN MATHEMAGICS
777 Left Bank Road Mullumbimby
NSW 2482 Australia

Ph: (02) 6684 4409
Email: jain@jainmathemagics.com
www.jainmathemagics.com

Videos & Books Available Online...
www.jainmathemagics.com
Copyright © Jain Mathemagics 2005

This Dvd will give the reader a total overview of Jain's Life's-Works covering the 4 distinct topics that he teaches world-wide:
1- Vedic Mathematics, 2- Magic Squares, 3- The Divine Phi Proportion, and 4- Three-Dimensional Geometry (of the 5 Platonic Solids).
Jain has subsequently extended this 4 armed curriculum to incorporate the 5th Element of Digital Compression now known as the Translation Of Number Into Art, or simply, **The Art Of Number**.

(Art by Jain, 1994, "Beings Of Air")

cHAPTER 2

ALGEBRAIC MUSINGS ON PHI LOVE

- ## ❖ TO UNI-PHI
- ## ❖ GLOBALIZATION ONE WORLD ORDER
- ## ❖ X=THE UNKNOWN=24=THE STARGATE
- ## ❖ Ponderings Upon the UNIT CIRCLE
- ## ❖ Phi Evolving Out of The Triune Unity

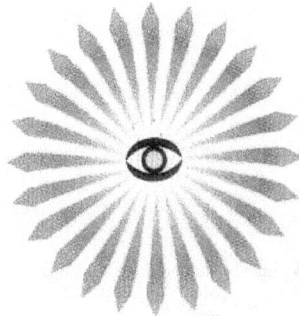

(logo for: "Inner Eye Publishing",
has the circle divided into 24 rays with an all seeing eye in the centre.)

pART 1
(INTERNAL DIVISION Of The UNIT CIRCLE)

The purpose of this article, as we learn two methods how to internally and externally divide the unit circle diameter into it's Golden Mean or Phi Ratios of .618033... (the Reciprocal of Phi whose symbol is **1/ɸ**) and 1.618033... (Phi whose symbol is **ɸ**) respectively, is to conclude that we Know Everything, a subject known as INTUITIVE MATHEMATICS.

We begin this Journey by constructing the blueprint of all creation, that which we call **THE UNIT CIRCLE**. It is always our starting point, the Point, the One, the God, the Bindu, the Primal Dot that extends it's radii to create Form.

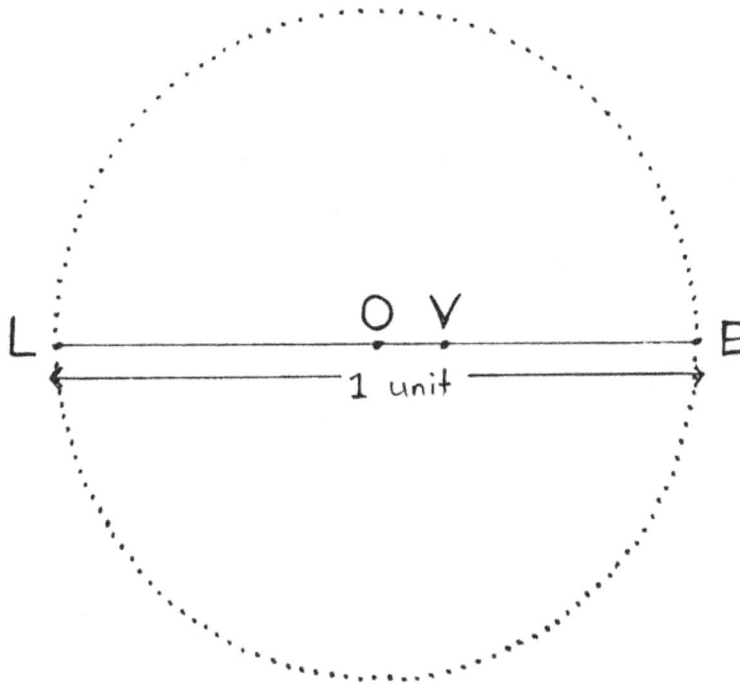

the UNIT CIRCLE
expressed by JAIN as L-O-V-E

Fig 1
"Creation's Love Diagram" by Jain (copyrighted 2008)
Depiction of the Unit Diagram with Letters
On how to Uni-Phi

The Diameter "**LE**" = 1 unit ("Le" in French, means "the" or "The"
the prefix that captures the Goddessence in all things or nouns.
The Centre is traditionally called The Origin and is expressed as
"**O**".
The Radius "OL' or "OE" = .5 or half.
Somewhere, the Golden Mean of LE will located precisely at "**V**", V
for Victory if you like, for V is what we are looking for, seeking,
questing for at the moment, we can only guess where V will be, but
any case, we need to make a mark and call it V, somewhere just
more than half, a fraction more than .5 is the Golden Mean division
of Unity Consciousness, somewhere about the range of .6, this is
nature's choice, not midpoints or binary, but a sexual tension just

beyond the stalemate of equal midpointed forces. Nature does not like equal spin as in 21 spirals to the left and 21 spirals to the right, but chooses the Fibonacci Sequence as optimized wavelengths that permit's infinite non-destructive travel, as in the sunflower's 21 to 34 counter-rotating fields in it's florets. This is the V Point, optimization of beauty, the living mathematics of nature, the Vpoint where the elbow bends, where the knee bends, where the digits of the fingers bend.

We can see this in Nature, but our purpose here today is to mathematically determine this Vpoint, using real ancient knowledge of algebra and quadratic equations. Don't be frightened yet of these words, algebra is indeed star language, allow me to show you how simply and elegantly this is done.

The answer of V, the Phi Ratio (1:1.618033...) is looking at us, staring us in the face, the face is all in the phi ratio, in fact on a higher level, we already know the answer, we have this Knowledge in our dna. We only need to perform a few fancy steps for the answer to be revealed, it's just a dance, an algorithm (an algorithm is like a cake recipe, we need to know the list of ingredients, that's all it is). But in this cake, there is something essential that is missing, one of the main ingredients is missing or actually it is unknown. And this is where the secret lies, in the **unknown**, the astrolgers 12th House, Neptune's nebulous domain of lost knowledge, but we are bringing it through, and the only way to do this is to give this Unknownness a name, to tame it as an Entity. Thus the ancient seers, or forbears called this concept of the Unknown "x" which is a letter of our English alphabet. In fact, they could of chosen any of the 26 letters, but why did they choose the 24th letter, 24 being one of the highly anointed or illuminated frequencies, vibration, of which I could write a whole encylopedia on.

Thus we begin here, that "x" = the 24th Letter, and understand that the secret of nature is what I call the Phi Code of 24 Infinitely repeating numbers in the Fibonacci Sequence whose sum is 108, it's a mantra, a chant that keeps repeating this pulse or hidden rhythm in all things called 108, 108, 108 ad infinitum.

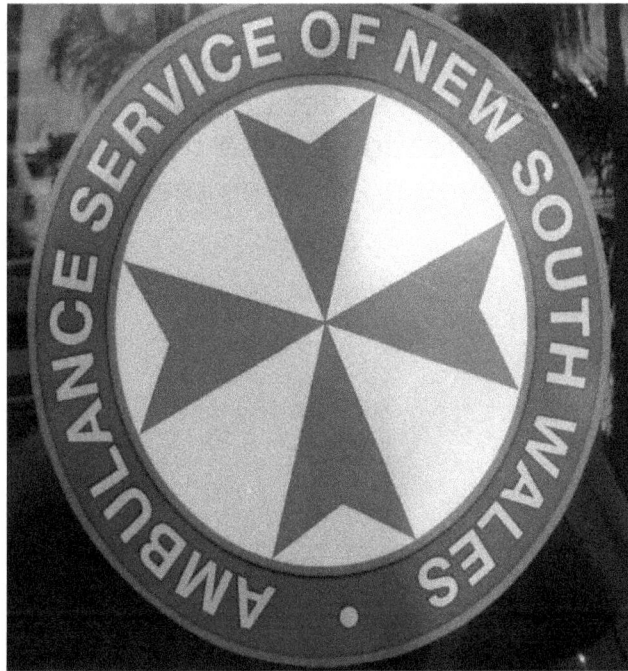

Fig 2
The Prime Number Cross Emblem of 24ness
known as the symbol for rescue healing:
Logo for the
The Ambulance Service Of New South Wales, Australia.

(Please read my other articles on the Phi Code 108 to get a deeper understanding of how Prime Numbers relate to Time Travel or the Physics Of Time Bending. That is why your clock or watch is divided into 24 segments or 24 hours, it is not an arbitrary choice of circle division, it is the Time Code, a veritable Star-Gate of **24** intelligent glyphs that permit's inter-dimensional space/time shape-shifting. Even the Queen of England wears this arcane symbol of "**X**" upon her royal cloak, it looks like the australian red ambulance symbol or maltese cross, a circle divided into 24 creates this emblem. Even the queen's military generals wear this same symbol which is technically called the Prime Number Cross, a mathematical diamond that captures the symmetry of all the Prime Numbers upon nested embedded wheels of 24 and suck all the energy of the universe out of it, as the Prime Numbers are the undivided **atoms of creation**, whilst you are taught at school that this is a nonsense entity, that there is no pattern or symmetry in the prime numbers, so don't go looking there **sheeple** as you might encounter a hidden dragon, so stay dumb and just get your grades at school and do what you are told, consume, get married, die, and meanwhile work at a job and pay off your mortgage and all is sweet until we orchestrate rising interest rates to keep you sweating on the treadmill so you don't have any free time to be revolting or even reading this dubious account).

HER MAJESTY THE QUEEN

This portrait of Her Majesty The Queen was painted by Mr. Leonard Boden in 1968. Her Majesty is shown in the robes of the Sovereign Head of the Order of St. John and wearing the Insignia of the Order which were made for Queen Victoria. On November 28th, 1968, Her Majesty visited St. John's Gate, Clerkenwell, where the portrait was on view for the first time.

Fig 3
The Prime Number Cross Emblem of 24ness
portrayed on the Queen of England's cloak
(for my indepth article on this, read in my next book:
The Book Of Phi, Volume 4, subtitled: Phi Code 1:
"Prime Number Cross, 24ness, and the Phi Code")

In the end, our purpose is to conclude that Algebra is a Star Language, and it is my task to at least switch you so that you don't fear Algebra, so stay with me as I turn you on.

Back to the unknown oddity called "x". The Algebra of "x" works because we are combining letters with numbers, by declaring that x=24 where "x" is a letter (Right Brain, language) and 24 = a "number" (left brain, logical). Thus this **Alpha-Numeric** phenomenon craftily joins the 2 hemispheres of the brain, creating **Whole-Brain Learning**". The English language does not do this, we have no association with numbers and letters, so we could say that we are technically lobotomized; in the old days, your name was associated to numbers, it was rich with analogies and symbolism, your name was also a picture... There are today several languages that still employ a "**numerological** value to alphabetical letters" where it is recognized sophisticatedly that a=1, b=2, c=3 etc. It is witnessed in the ancient Greek language, Hebrew and Sanskrit. One day our children will be protesting on the streets for the reinstallment of such knowledge. It was recorded, in the story about the **Tower of Babel** in biblical Babylon, when the whole world spoke One Language or should I say, **spone** language. I am preparing the countries of this world to gracefully return to this clever and intelligent One World Language, **One World Mathematics**, One World Currency. It's not good or bad, it just Is. I know what u r thinking, "Oh Jain has lost it, he is like a **666 FOX** guilefully steering us towards a sinister controlling **One World Order**, charming us with his Phi Code and Anointed Number 24 Musings calling it **Fixed Ancient Design** that our children must learn and remember, conning us, dazzling us as concerned parents to bring reform and change. Well, my response to that would be that there is no doubt or question regarding this enchanting material as the Truth Of Mathematics and the harbinger of **Unity Consciousness**. It is undisputable. The Circle has a diameter of One, oneness, you can't argue with this, here in Fig 1 above, we call it LE=1. Wait till you see what unfolds from this diagram, like wow. Yes, I sincerely believe that these unity diametric harmonics will unite all of our separated countries that are clashing and bitching with one another, because they don't understand this wholeness, this LE=1, they never learnt any of this at school. So support me, as a **mathematical futurist**, not destory nor destroy me. Yes, this crytpic mathematics has led me to dreamy thoughts of Globalization, where we all indulge in **One Currency**, One common train track gauge, no borders, yet maintain our dear ethnicities and cultures. You know surely that the one reason why we can't communicate with one another is because we are not yet globalized. Thanks though to the Internet, which we netizens love, we are already supporting this **Globalization**, so stop resisting it. We demand a One World Mathematics [of which I have already written the curriculum for in my series of books: JAIN MATHEMAGICS CURRICULUM FOR THE GLOBAL SCHOOL]; it's not a "gonna be", it's a reality happening as you read this.

Let us return to the Unknown "x" in the Unit Circle. The Definition of the Golden Mean **Fractality** is to **make the Inside the same as the Outside**. So we desire to locate "V", the Mystic Cut, in Fig 1, where:

LE / LV = LV / VE

(The symbol "/" of the "Forward Stroke" means "Divided By") and we could have written it like this shown below:

$$\frac{LE}{LV} = \frac{LV}{VE}$$

ie: the biggest whole segment LE=1 when divided by the larger segment or proportion LV is in the same fractal ratio or division as this larger proportion LV divided by the smaller proportion VE.

$$\frac{Whole}{Larger} : \frac{Larger}{Smallest}$$

Really, this most ancient puzzle of how to divide the line into 3 proportions (whole, larger, smallest) is indeed a **Trinity Relationship**. That is why the Christians adored the Golden Mean, as it symbolized to them their "Father/Son/Holy Ghost".

Algebra requires that you put your thinking cap on. (In the old days, the "**Dunce's Hat**" was a cone shaped hat that stimulated the brain, to wake up a dreamy and lazy student. The cone shaped hat was based on the Phi Ratio, where the head diameter was = to 1, the slope height was Phi = 1.618033... and the height of the hat

was

the Five Fishes
in the Bermuda Golden Triangle

therefore The Square Root Of Phi = 1.272...

Fig 4
The conical Dunce's Hat
employing the light harmonics of
Unity, Phi and the Square Root of Phi.
The triangular shape around the head
is based on the mathematics of the Pentacle
(Painting by Jain in 1982 in the Torres Strait Islands
known as "Haha Aha and the 5 Fishes").

Up till now, we know two distances, which will help us define the 3rd distance. That is, we know that
1)- diameter LE = 1
2)- we crafted that LV="x", which is the larger segment, that we have called the **Unknown**, (and this will permit us to discover the smaller segment)
3)- VE=?

This is shown below in Fig 5:

Fig 5
**The 2 of the 3 known distances
established in our LOVE Diagram**

At this stage, observing this diagram and not reading ahead, can
you tell me what the distance of VE is?
Any young teenager can look at this data and understand logically
that this smaller distance VE must be LE minus LV
ie: VE = LE – LV
     VE = 1 minus "x"

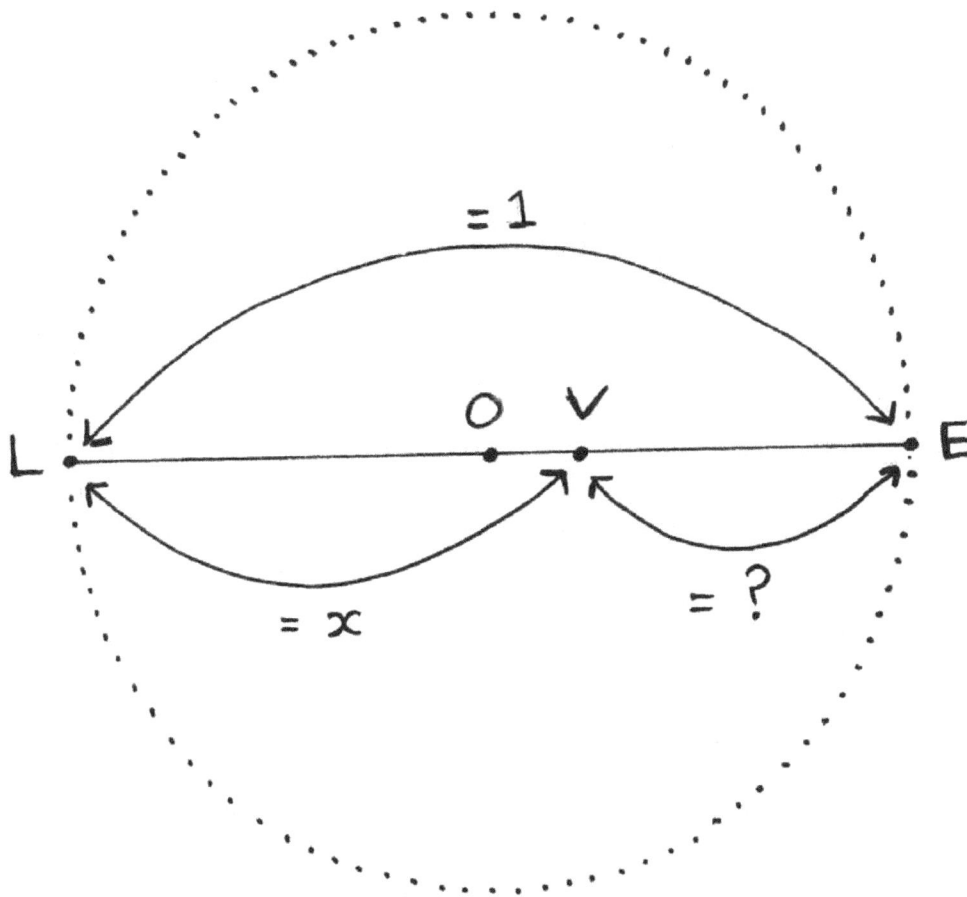

Unit Circle trinitized into 3 segments

VE = 1 – "x"

nb: In this article, the symbol for multiplication "x" will not be used
anywhere in the text, so when you see either x or "x" it refers to
the 24th letter of our alphabet being "x".

We are on our way know, knowing that VE = 1– "x"
and since it is a bit like a wild panther caught in our cage, we need to contain this rare catch, this powerful force spawned from mother earth, by putting parentheses around it, and declaring that our catch is labelled as:

VE = (1– x).

We've done it, we are now half way to getting towards a cosmic and mathematically engineered result. It's simple and brilliant.
At school, you may have learnt how to **Cross Multiply** when adding fractions like ¼ + 1/3

= (1 times 3) + (1 times 4)
              3 times 4

What we did on the top line [(1 times 3) + (1 times 4)] was to utilize the shape of a big "X" which we call Cross Multiplication. The Indian Vedic Mathematicians from 2,000 years ago used this Cross Multiplication to achieve rapid mental calculation. It was given a name "**Vertically and Crosswise**" as one of 16 possible Sutras that could solve all mathematical problems either mentally or in one swift line of written calculation. This word they used for the 16 cryptic word formulae is "**Sutra**" literally meaning "Thread of Knowledge" or a timeless Law of Numbers. (You can study more about this in my other books).
Thus to solve our Puzzle at hand, we will employ this Crosswise Sutra to solve

$$\frac{LE}{LV} = \frac{LV}{VE}$$

and by substituting the values LE=1, LV="x" and VE=(1–x)
we arrive to this grand and mysterious mathematical plateau:

$$\frac{1}{x} = \frac{x}{1-x}$$

This is what nerds like me call an elegant equation.
Now don't be scared on what it morphs to, it's still the same thing, it's just simply evolving to a state where we equate it to **zero**, so hang in there, while we now Cross Multiply both sides, in a sense, to stretch it out or expand the equation so that we can jiggle around with it's terms.
(Generally, to cross multiply say

$$\frac{A}{B} = \frac{C}{D}$$

the result is A times D = C times B).
giving us:
1 times (1−x) = "x" times "x"
which can be simplified to:

$1-x = x^2$

To tidy up this **algebraic equation** the ancients said that whatever we do to one side of the equation, to maintain equanimity or balance, we must do the same to the other side.
To rearrange it so that all the data of both the letters and numbers are conveniently placed on one side of the "=" sign, we are going to subtract (1−x) from the Left Hand Side (LHS) so that the (LHS) is zeroed out. Having done this, we need to do the same subtraction of (1−x) from the Right Hand Side (RHS), giving:

$(1-x) - (1-x) = x^2 - (1-x)$

$0 = x^2 - 1 - - x$ (Two minuses being multiplied give a positive)

$0 = x^2 - 1 + x$

Conventionally, we rearrange this again ut (such that) the "0" is on the RHS and write as a final expression:

$x^2 + x - 1 = 0$

(This algebraic equation can be read as "something x multiplied by itself, plus that something and lessened by 1 equals nothing or zero!

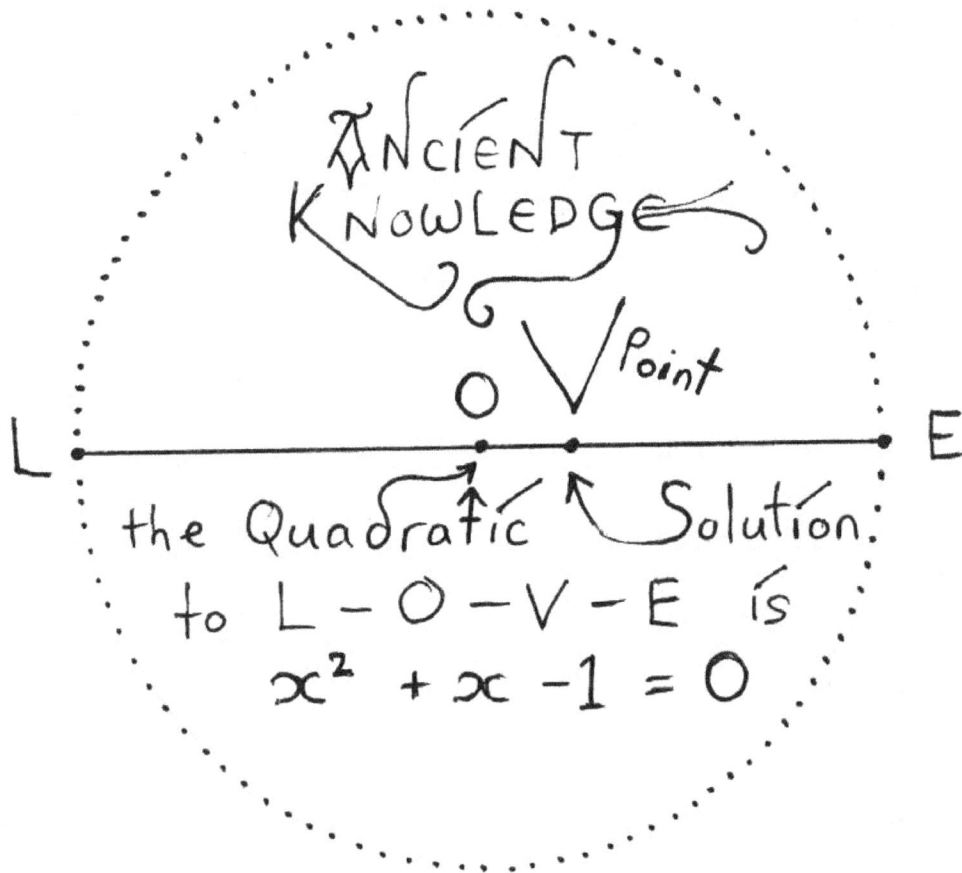

Ancient Knowledge

O · V Point

L ——————————— E

the Quadratic Solution
to L – O – V – E is
$x^2 + x - 1 = 0$

Fig 6

The Solution to L-O-V-E is $x^2 + x - 1 = 0$

I know that this is starting to look scary, to see this raw wild incomprehensible gibberish being thrown at you, yet it is a clue to how the universe operates, it's the real treasure that Indiana Jones was looking for all this time, it's that ethereal invisible essence perhaps from other star systems which is embedded in the **hypothalmus** cave of brahma of those crystal skulls.

Before we extract the .618 golden meaning from inside the guts of this equation, I would like to first demystify what an algebraic equation is. This part is really embarrassing, but I must admit it here, as it may also be your story or experience at school and may help you, but at Kirrawee High School, studying quadratic equations like the above (as "quad" means four, or 4 parts of the equation) like say $2x^2 + x + 6 = 0$ is a quadratic equation with 4 parts, meant nothing to me at high school, as the teacher never explained it to me. I was so caught up or brainwashed in achieving high marks to get to university to study medicine, that I merely learnt this factory style mathematics, like a robot, rote learning, getting answers right, but with no soul. We were told to factorize things like $2x^2 + x + 6 = 0$. But so what, it was more out of fear to get it right and being overwhelmed by it's meaning, that I never understood for those 6 years at high school that the number 216 is really $2x^2 + x + 6$ in Base 10. You know how our numbers are derived from the Decimal Place Value System of Unit's, Tens, Hundreds, Thousands etc, well all our numbers can be expressed in this funny format like $x^3 + 6x^2 + x + 8 = 0$ is really another strange way of say 1,618, like der, like why didn't my maths teacher ever explain that to me, I guess studying all those logarithms and simultaneous equations kept me brain dead. Don't get me wrong, logarithms are also star language, I get excited just by seeing that word. How ridiculously embarrassing that I a genius went thru skool as a victim of the state, mindless like a fenced cow instituted with over thousands of other young sheeple, and just didn't get the meaning of algebra. Though, through all of this, I do get it, 30 years later, get what I mean.

OK, enuf waffling on, lets finish this story of how .618033... is fractally squeezed out of the trinitized unit circle. We're nine parts out of ten nearly there, approaching that adorable **Vpoint**, as we quested like Indiana Jones and discovered a clue in the jungle of our consciousness a weird and mysteriously mystical formula:
$x^2 + x - 1 = 0$
so close to extracting this .618033... golden mean proportion. The final step is derived from another complex looking mathematical formula that solves Quadratic Equations, and is known as the Quadratic Formula (it is another lesson on how this was derived, but for now just accept and learn this fact:

$$\text{for } ax^2 + bx + c = 0$$
$$\text{the Quadratic Formula}$$
$$x = \frac{-b \pm \sqrt{b^2 - 4ac}}{2a}$$

Fig 7
The Quadratic Formula that is the last step required to determine the roots of $x^2 + x - 1 = 0$ that will generate the Golden Mean of our Unit Circle

(The ancient Vedic Mathematicians were able to bypass this complex looking formula and actually determine the roots of the equation by mental means only, without use of pen or paper)!

I am going to re-type the Quadratic Formula shown above in Fig 7 using only one line to help spell out what it means, so we can get to the solution or answer that = .618033...

x = minus b plus or minus the square root of b squared minus 4 ac all divided by 2a
This Quadratic Formula gives the roots of any general quadratic equation in the form of:
$ax^2 + bx + c = 0$
which means we need to know the numerical values of the 3 letters a, b and c which are called co-efficients.
So lets look again at our derived quadratic equation:
$x^2 + x - 1 = 0$
and write down these coefficient values now where
a=1, b=1 and c=−1

(because $x^2 + x - 1$ is really the same as $1x^2 + 1x - 1$)
and substitute those 3 values of a=1, b=1 and c=−1 into the
Quadratic Formula in Fig 7 that will give the final dual solution of:

Minus 1 plus or minus the Square Root of 5 all divided by 2

$$\text{Phi } \phi = \frac{-1 \pm \sqrt{5}}{2}$$

$$\frac{1}{\phi} = \frac{-1 + \sqrt{5}}{2} \qquad \phi = \frac{-1 - \sqrt{5}}{2}$$

$$= + \cdot 618... \qquad = -1.618...$$

Fig 8
The generalized Phi solution,
but really there are two roots,
this positive one, but also one in the imaginary world
symbolized by the negative axis and numbers.

Here are the 2 roots, one is positive, and the other negative.

phi = (−1 + Root 5) ÷ 2
   = (−1 + 2.236067978...) ÷ 2
   = (1.236067978...) ÷ 2
   = **.618033988...**

(traditionally, **Phi** refers to the value 1.618033988,
and the Reciprocal of Phi which is 1 ÷ Phi = .618033988 and is
expressed as **phi** with a small initial **p** as shown above).

Phi = (−1 − Root 5) ÷ 2
   = (−1 − 2.236067978...) ÷ 2
   = (−3.236067978...) ÷ 2
   = −1.**618033988...**

Thus the solution as to where the Golden Mean Division of the Unit Circle, along the diameter of LE=1 is precisely the length of LV being equal to .618033...
(the 3 dots at the end of this number indicates that the decimal continues forever without any known repetition or symmetry, an infinitely travelling decimal that looks like
**.61803398874989484820458683436563811772030917980
57628621354486227052604628189024497072072 0418939
11374847540880753868917521266338622235369 3179318
00607667263544333890865959395829056383226 6131992
82902678806752087668925017116962070322210 4**...

Thus this was the elementary lesson how to Trinitize the Unit Circle by application of what I call Internal Division, as all the mathematics was done inside of the unit circle, which is why our answer of .618033... had to lie from 0 to 1.

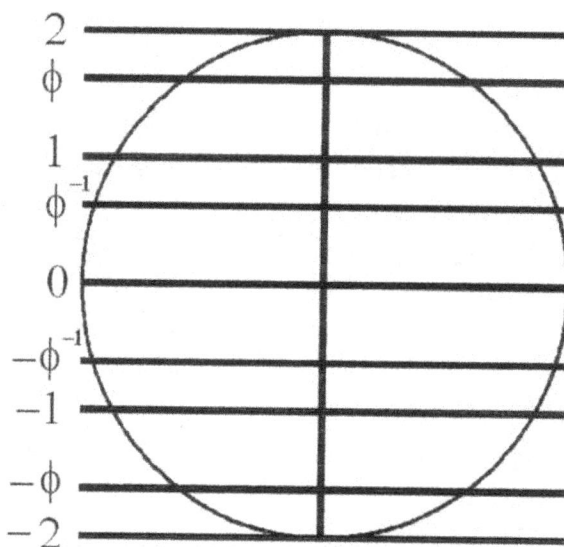

Fig 9
(Unit Circle Divided in Phi Ratio Harmonics or Divisions)

pART 2

(EXTERNAL DIVISION of the UNIT CIRCLE)

Have a look at this diagram Fig 8 and predict what is asked of you.
How does it differ from Fig 1, our starting point.

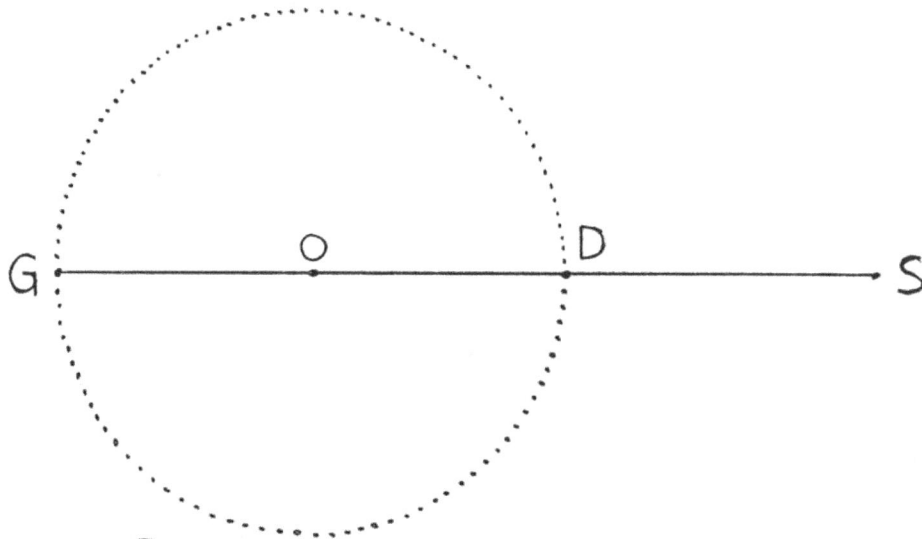

Fig   **9**

Eternal Phi Division of Unit Circle

The External Division of the Unit Circle

The Advanced Level to this research, and is your homework, is to
consider a complementary diagram to the Internal Division of the
Unit Circle which is suction towards the Centre as in a **centripetal**
force, or is the principle of **implosion**, and investigate "The
External Division of the Unit Circle" as in a **centrifugal** force, that
which is the principle of explosion, where the diameter GOD = 1
and DS is the Unknown value called "x" and must have a value for
GS greater than one, and we know that the answer is GS = Phi =
1.618033988...
Your Homework is to determine the cross multiplication
mathematics that sets up the quadratic equation.
I know yar not gonna do it,
so here it is anyway:

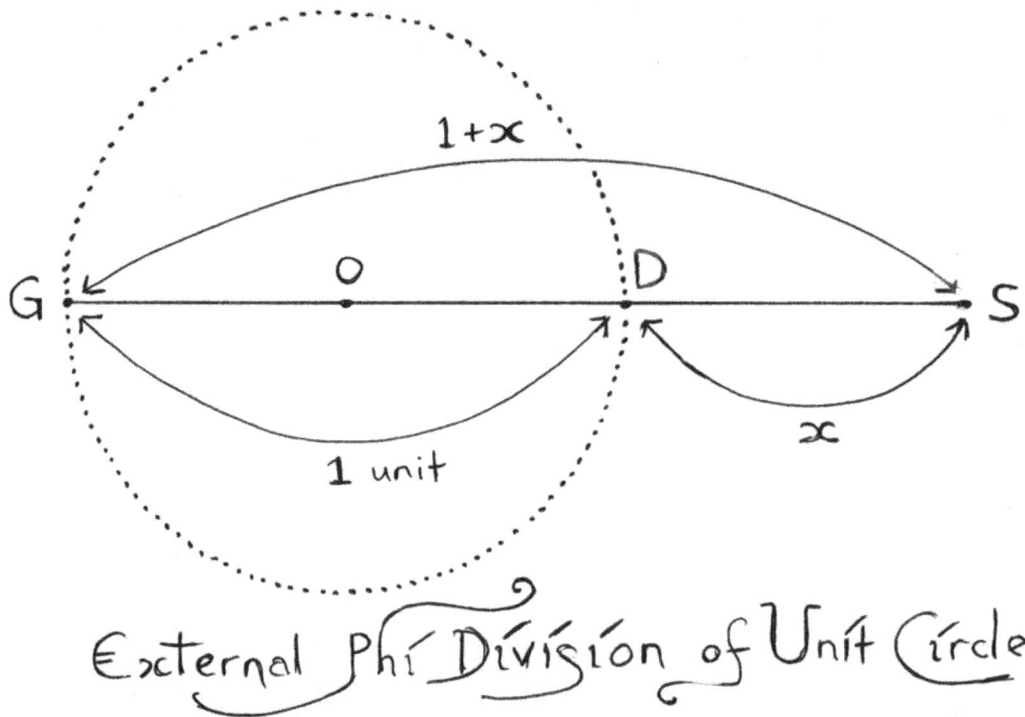

1+x

G ⟵ O ⟶ D ⟶ S

1 unit

x

Eaxternal Phí Division of Unit Circle

Fig 10
The External Division of the Unit Circle
showing the 3 proportions of diameter GD=1,
the smaller and external Unknown segment "x" = DS,
and the whole line GS = Phi = 1.618033...

$$\frac{GS}{GD} = \frac{GD}{DS}$$

$$\frac{1+x}{1} = \frac{1}{x}$$

$x(1+x) = 1$
$x + x^2 = 1$
$x^2 + x - 1 = 0$

$x = (-1 + \text{Root } 5) \div 2$

which gives the same answer as before
external length DS = phi = .618033988...
this gives the value of the whole length GS to

= Phi
= 1.618033988...

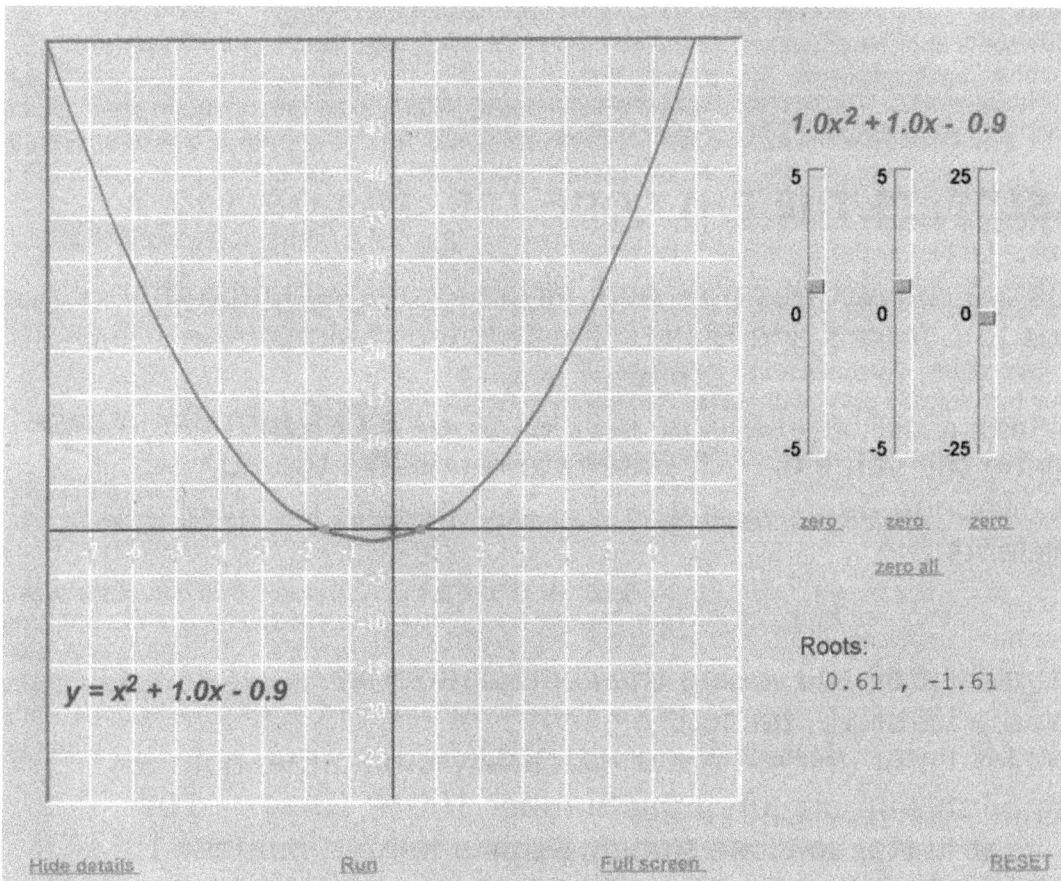

$1.0x^2 + 1.0x - 0.9$

Roots:
0.61 , -1.61

$y = x^2 + 1.0x - 0.9$

Hide details   Run   Full screen   RESET

Fig 11
The Graphical Derivation of Phi
The graph of y = x^2 + x -1
(which is of the form y = x^2 + bx +c)
(source of this image is taken from the Internet:
http://www.mathopenref.com/quadraticexplorer.html)

The two roots of this now famous quadratic equation y = x^2 + x -1
is found by observing where the parabolic-like curve intersects the
horizontal "x"-axis. In the diagram of Fig 11, two small blue dots
have been plotted on the "x" axis to show this dual solution.
Close observation shows very graphically that the two roots are
+ve .618033988... and **−ve 1.618033988...**

(nb: when I found this **Quadratic Function Explorer Graph** on a website

[http://www.mathopenref.com/quadraticexplorer.html] it would not allow me to precisely graph $y = x^2 + x - 1$
but would only allow me to graph $y = x^2 + x - .9$ which is very very close, but not exact. No matter though, it serves it's purpose to illustrate that the roots of phi lie on the horizontal axis).
See also: http://www.vashti.net/mceinc/rgphigra.htm for other Graphical Representations of Phi.

An ASIDE: IS THE PHI CODE 108, 109 OR 117 ?

How do we convert this now popular quadratic equation that belongs to Nature's and Biology's esteemed realms:

$$y = x^2 + x - 1$$

As explained earlier in this article, we rewrite this formula in base 10, giving the value of "x" as ten, to represent the number in terms of hundreds, tens and unit's. This is done by observing it's 3 co-efficients:

$$y = 1x^2 + 1x - 1$$

We have 1 lot of 100, being the coefficient of x^2.
We have 1 lot of 10, being the coefficient of "x".
We have 1 lot of unit's, here it is actually a negative number "–1."
Adding all this up procures the number $100 + 10 - 1 = $ **109**.
Oohh, that hurts, so close to being this anointed number 108.
But this predicament is where discoveries are made, through all **anomalies**, one must travel down that rabbit hole to see where it will lead. Like true science, we can not fudge true mathematics. No matter how much we wish it summed to 108, that would have been a major mathematical discovery, that the holy Sri Vedic 108 was always disguised in phi's quadratic expression.

As we conclude this article soon, I would like to therefore plug the importance of this earthly world having a bona fide Numerical Dictionary of Anointed Numbers, a heavenly and very visual number dictionary that restores Pythagorean lost knowledge. That is why I call it "**Harmonic Stairway**", and envision a team of people scanning and printing and filing all important harmonics and number codes and patterns to be made available in our online university and **Temple of Mathematics**.
For example, right now, we stumbled upon this number 109, and

need urgently to open up Jain's Dictionary Of Harmonic Numbers and check to see whether or not there are any intelligent correlations about the number 109 and other numerical entities. As you can see, I have devoted half my life to passionate exploration of this Phi Code 108.

What if the Phi Code is more than 108, it is possible that the Phi Code is embodying some other digital principle that we are not even aware of.

Let's examine it again, as there is more in it that you would suspect:

Jain's 108 Phi Code: an Infinitely Repeating 24 Pattern
Based on the Compression of the Fibonacci Numbers
into Single Digits

1	1	2	3	5	8	4	3	7	1	8	9	8	8	7	6	4	1	5	6	2	8	1	9

PHI CODE of 12 COMPLEMENTARY PAIRS OF 9

1st Set of 12 Numbers   1 1 2 3 5 8 4 3 7 1 8 **9**

2nd Set of 12 Numbers   8 8 7 6 4 1 5 6 2 8 1 **9**

Fig 12
**The Infinitely repeating 24 Pattern
shown as two lines of 12
Highlighting the intriguing final pair of double Nines.
What does it really mean?**

Notice that overall, there are 12 Pairs of 9 which = 108, except for the last pair that is a double 9 or a double Pair, which seems to act as a bridge or a bond for the infinitely continuous 108 code.

It appears, that the ancient secret goes like this:

108 – 9 – 108 – 9 – 108 – 9 or tabulated as:

Infinite "108 -9" Phi Code Sequence								
108	9	108	9	108	9	108	9	108

Fig 12

The Infinitely Recursive 108 – 9 Phi Code Sequence

So we can see that this Phi Code
is a continuous infinity of **108 – 9.**
Could this double nine pair reduce to 1 to make the 108+1=**109** ?!
Really, all the numbers in Fig 12, if you counted them including the
pair of double nines, adds up 108+9 =**117** which is interestingly a
prime number and also warrants research in the Harmonic
Stairway Dictionary.
Thus how do we conclude, Is the Phi Code 108 or 109 or 117 ?
At this stage in our Journey, it is not the outcome, it's not
important which number it is, that which is important is the
recognition that the Process Is The Goal, the doing, the current
questing, the ability to manifest world institutions that are
contributing to this Journey, to be able to leave a legacy behind for
the incoming race of crystal indigo star children.

Regards **Jain 108**
1-10-2008
Mullumbimby Creek, Australia

pART 3

ON THE CURIOUS NATURE OF 109

The following information is taken from www.phinest.com showing the **curious nature of 109** based on the fibonacci numbers:

The reciprocal of 109 is also based on the Fibonacci series, forwards and backwards

Here's another curiousity involving the number 109, discovered and contributed (10/20/2003) by Rick Toews.

1/109 is a repeating decimal fraction with 108 characters:

.00917431192660550458715596330275229357798
16513761467889908256880733944954128440366
972477064220183486238**53211**

Fig 13

The 108 decimalized digits in the Reciprocal of 109

You can see, in bold and underlined above, the beginning of the Fibonacci sequence in **the LAST 6 digits** of the decimal equivalent of 1/109, appearing in **REVERSE** order starting from the END of the decimal. (i.e., 0,1,1,2,3,5,8 appears as ...853211)

If you take each Fibonacci number, divide it by 10 raised to the power of 109 MINUS it's position in the Fibonacci sequence (starting with 0) and add them all together, you get the reciprocal of 109.

To understand this better, look at the numbers below, especially on the far right hand side where the whole Fibonacci Numbers are being tabulated and added together:

1 / 109 =

.0091743119266055045871559633027522935779816513761467889908256880733944954128440366972477064220183486238**53211**

---

0 / (10 ^ 109)   + ...0000000000000000 +

1 / (10 ^ 108)   + ...00000000000001 +

1 / (10 ^ 107)   + ...0000000000001 +

2 / (10 ^ 106)   + ...000000000002 +

3 / (10 ^ 105)   + ...00000000003 +

5 / (10 ^ 104)   + ...0000000005 +

8 / (10 ^ 103)   + ...000000008 +

13 / (10 ^ 102) + ...00000013 +

21 / (10 ^ 101) + ...0000021 +

34 / (10 ^ 100) + ...000034 +

55 / (10 ^ 99) + ...00055 +

89 / (10 ^ 98) + ...0089 +

144 / (10 ^ 97) + ...144 +

233 / (10 ^ 97) + ...33 +

377 / (10 ^ 97) +

Fig 13a

The addition of the Fibonacci Numbers in a curious manner,
shown on the far right hand side in bold and underlined,
generate the Reciprocal of 109

Lastly, here's one more curiousity involving the number 109.

It is also taken from the internet source already acknowledged above.

If you take each Fibonacci number, divide it by 10 raised to the power of its position in the Fibonacci Sequence and **add and subtract** each alternate term together, you get .00917431... again, the reciprocal of 109.
Here it is tabulated below:

1 / 109 = 0.00917431...=

$$0.\underline{0} +$$

$$0 / (10 \wedge \underline{1}) + 0.01 -$$
$$1 / (10 \wedge \underline{2}) - 0.00\underline{1} +$$
$$1 / (10 \wedge \underline{3}) + 0.000\underline{2} -$$
$$2 / (10 \wedge \underline{4}) - 0.0000\underline{3} +$$
$$3 / (10 \wedge \underline{5}) + 0.00000\underline{5} -$$
$$5 / (10 \wedge \underline{6}) - 0.0000000\underline{8} +$$
$$8 / (10 \wedge \underline{7}) + 0.000000\underline{13} -$$
$$13 / (10 \wedge \underline{8}) - 0.0000000\underline{21} +$$
$$21 / (10 \wedge \underline{9}) +$$

Fig 13b

The addition and subtraction of the Fibonacci Numbers in a curious manner, shown on the far right hand side in bold and underlined, generate the Reciprocal of 109.

aBOUT tHE aUTHOR

Jain 108 is a self-taught mathematical futurist,
a numerical nomad and a digital genius.
Some of his colleagues describe him as a mathematical monk.
He hails from a quaint Sydney suburb Kirrawee, born in Australia
from centuries if not thousands of years of full-blood Phoenician
inter-marrying stock, ne 1957.
His great grandfather in a remote village of Lebanon was a priest
from which his mother derived her maiden surname "Khoury".

**Collin Nicholas Saad aka Jain has had likeable and resonant
long-time associations with Mar Charbel, a humble Lebanese
priest from circa 100 years ago now ordained officially as a
saint of the Roman Catholic Church. But we suspect that Jain
hails from a noveau kind of ministry, one that he has termed
in his voluminous writings as "Sacerdos Libri Naturae"
apparently = "a Priest of the Book of Nature".**

BINARY V e r s u s PHI CODE

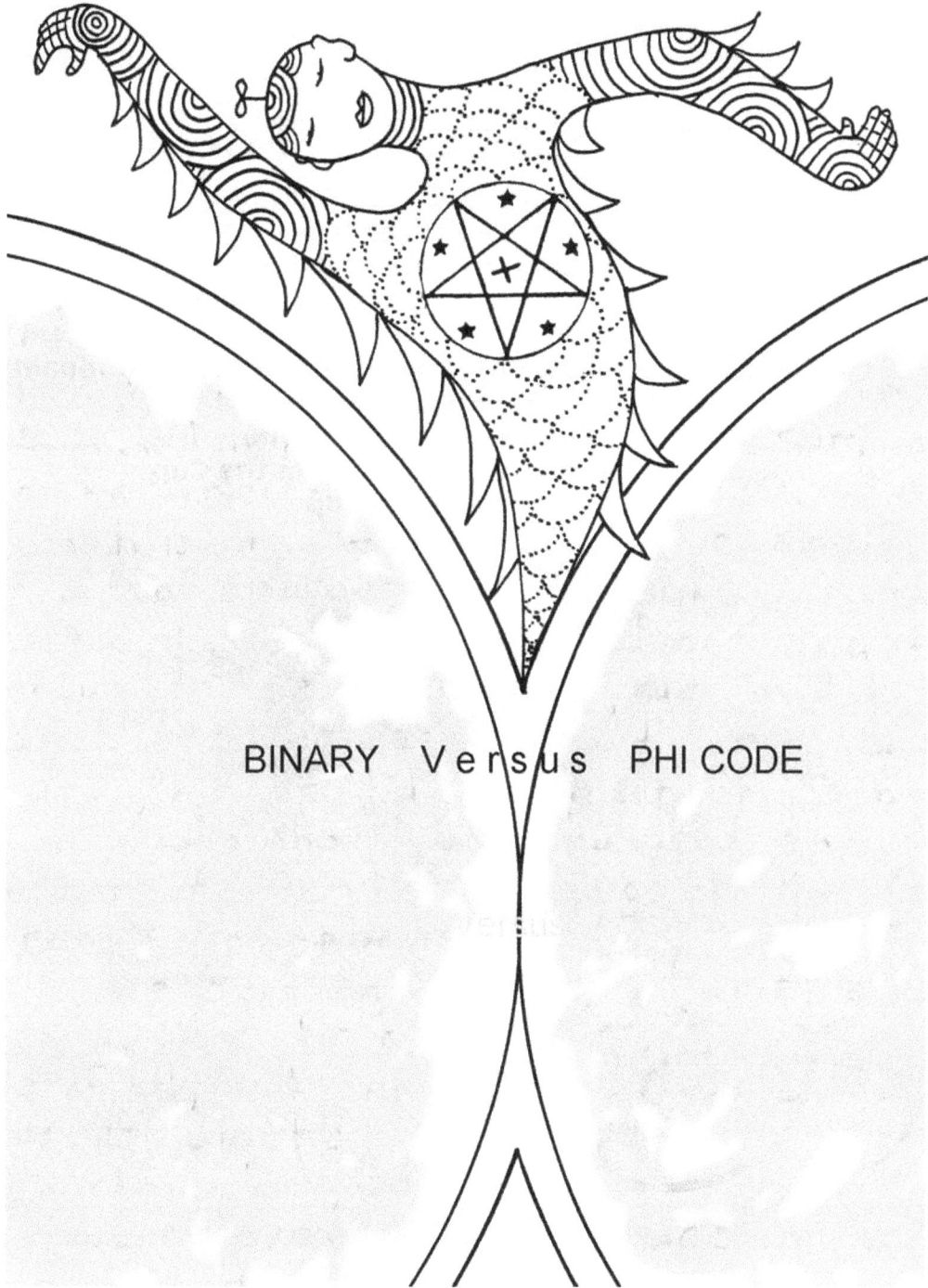

(Art by Jain, 1994 "Ether Dancer")

cHAPTER 3

The Binary Code Vs The Phi Code

dedicated to
the ENGRAILED ONES
those who live by and are modeled by
the most ergonomic and self-organized Phi Phi-Nest
who have awakened from the apparent dream of duality.
There is no Versus;
Binary and Phi are One
when the Cup flows within the Cup
within the Cup

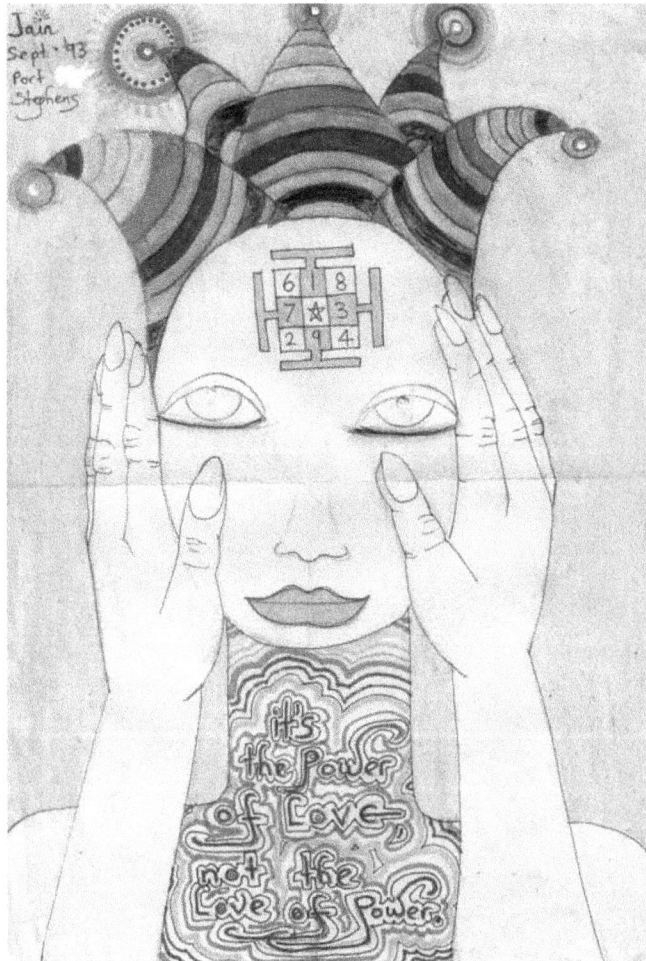

(The Art Of Jain, 1993)

Binary Vs Phi Code
Includes:

* Definition of **Binary**
* **Equal Spin** Vs **Phi Spin**
* **Geometric Proof** that the Rings of the Binary Code
  (1:2:4:8) are indeed in a Phi relationship!
* The Binary Code in **Cellular Division**
* Chart of "**Cosmic Vibrations**" mapping the **Powers Of Two** that
  define the **Electromagnetic Spectrum** etc
* Binary Versus Phi in tabulated form
* The **Doubling Sequence** And The **Mathematical Origin** Of The
  VW Symbol!
* Plot the Doubling Sequence: **1   2   4   8   7   5**
  on the **9 Point Circle**.
* Examination of the Pattern in the **Pairs of the 9 Times Table**.
* Notes on the **Archimedean Solid** called the **Cuboctahedron**.
* How the **24 Edges** of the **Cuboctahedron,** and the **24
  Equiangular Triangles** of the **Star Tetrahedron** relates to the
  Phi Codes 108.
* Creation of the Cuboctahedron's **Dual Polyhedron** called
  The **Rhombic Dodecahedron** having **12 Diamonds faces**.

Art Of Jain 1999

What does binary mean?

First of all, in contrast, our Base Ten or Decimal System ingeniously uses the numbers from 0 to 9.

Without the concept of a Zero, we would all still be living in caves. The **invention of Zero** is perhaps the greatest gift to this modern world, for without it, we could not have computers!

A base 2 system means that all the place values are a power of 2. In a base 2 system there can only be a "**1**" or a "**0**" in any given place, (On/Off, Yes/No) based on the following structure:

128	64	32	16	8	4	2	1
2^7	2^6	2^5	2^4	2^3	2^2	21	2^0

Thus the number **216** would be expressed as:

128	64	32	16	8	4	2	1
2^7	2^6	2^5	2^4	2^3	2^2	2^1	2^0
1	1	0	1	1	0	0	0

Thus, 216 written in Base 2, is therefore **11011000** which means that it is the sum of : 128 + 64 + 16 + 8.

Here comes a brief discourse and teaching on the distinction between the two most important codes: **The Binary Code** Versus **The Phi Code.**

Ultimately, I am saying that we are neither this or neither that, we are both, and this can be proven geometrically and dramatically in one beautiful diagram that shows how the concentric rings of the binary code, as they expand from one to two to four etc, have a phi ratioed connection. It's beautiful, and teaches that we are all Codes or Sequences. **Quantum Mind** or One Mind, **ONEarth.**

It's only the male mind that wants to do this, to pull things apart, analyze, investigate, whereas a women, she already knows all the sacred mysteries, even all the mathematics is in her highly inherited **intuitive** grasp.

BINARY CODE: 1 - 2 - 4 - 16 -32- 64 - 128 - 256 - 512 etc
(we were this binary code, from the original cell in our
CosmoGenesis: father-mother cell was **The One** that became **The
Two** that became **The 4** (whose 4 centres make the **Tetrahedron**)
that became **The 8** (whose 8 centres became the **Star
Tetrahedron** or Cube) whose final **mitotic division** of endless
halving collapsed it's superb geometry at **512** cell division and
became The **Tube Torus Doughnut**! Yes we were once this ring
shape having two holes, one for the **Mouth** that received the
universe, and one for the **Anus** that released the universe.....and
from this primal **Shape** we unfurled into a fern-like entity whose
overall shape was the golden mean spiral up to the point now
where every part of our being, like where the elbow bends, is in the
divine phi 1:1.618 proportion....

This is binary code: when the **One became the Two**, which
symbolizes separation from God...... and is about Technology, the
Machine or **Borg-Hive Consciousness**, Electronics, the "**Zero and
One**" of computer digital language.

It's about **midpoints**, halving, point 5 ratio (.5) or halving of the
whole... It's symbol is the **Equal Spin** of the **Sahasrara** or Crown
Chakra, where **1000 spirals** spin clockwise superimposed by 1000
spirals spinning counter-clockwise. Whatever number of spirals, it is
balanced and equal in it's **counter-rotating fields**.

Of note here, we can state that Nature does not do this, (Nature
does 21:34 as in the **Sunflower Map**).

It is my observation here that the masters of ancient times who
encoded this Equal Spin for the Crown Chakra, did it as a deliberate
ploy to inhibit **Time Travel**, to create a ceiling, a barrier into the
unlimited nature of the Phi Code.

There are some **Brahmin** (Hindu Priests) and Yoga Masters secretly
teaching this madness of Equal Spin in the Crown Chakra to
actually control them or limit their Consciousness. Be very wary of
any teachers promulgating Equal Spin. Ask them why they are not
celebrating the unlimited aesthetics of the Phi Spiral Spin Ratio as
in the Sunflower, which is our true barometer of spiritual
awareness, when we are attuned to the flowers.

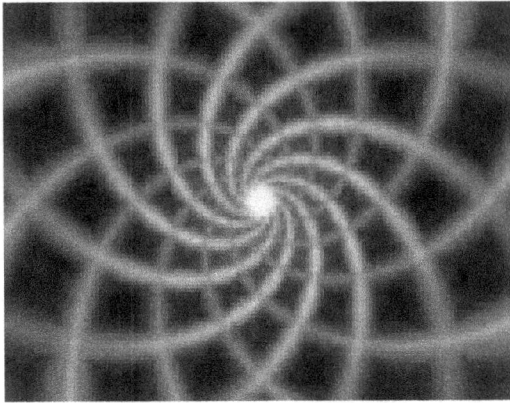

Fig 1a
"Equal Spin" 10 spirals going one way,
and 10 going the other way.
This is not Nature.

Have a look at Fig 1a above, and Fig 1b below, they are really both the same phenomena.

The diagram of Fig 1b shows what we call "**Equal Spin**" **36 spirals** going one way, and 36 going the other way.

What this does, when it is visualized at the top or crown chakra is to effectively start the motor, it merely turns it on, agitates energy, gets it spinning, but it does not go anywhere, it is like a car whose motor is turned on, but it is just idling. It does not travel. Whereas, in the next section, you will see what nature does, the 21:34 spin code which permit's travel through the dimensions.

In summary, Binary Code is efficient, but it is unlimited.

Phi Code is Nature's choice, it is unlimited.

We are examining here the ancient and lost science of what we call "**Counter-Rotating Fields**". Many hidden experiments like the infamous "**Philadelphia Experiment**" were using this concept of spin based on the Fibonacci Numbers to make battleships invisible. It's a very controversial and dark topic at the moment and we will not go there, for now. Ultimately, it did not work, as the pure intent was not pure, rather it was more militaristic and power hunger. The principle was correct, **to copy and comprehend Nature**, to utilize this apparent counter-rotating field, not as equal spin but according to the 21:34 code of the sunflower; one direction of the spin is the mental body, and other direction of the spin is the emotional body; how to get them in sync is the key to **invisibility**.

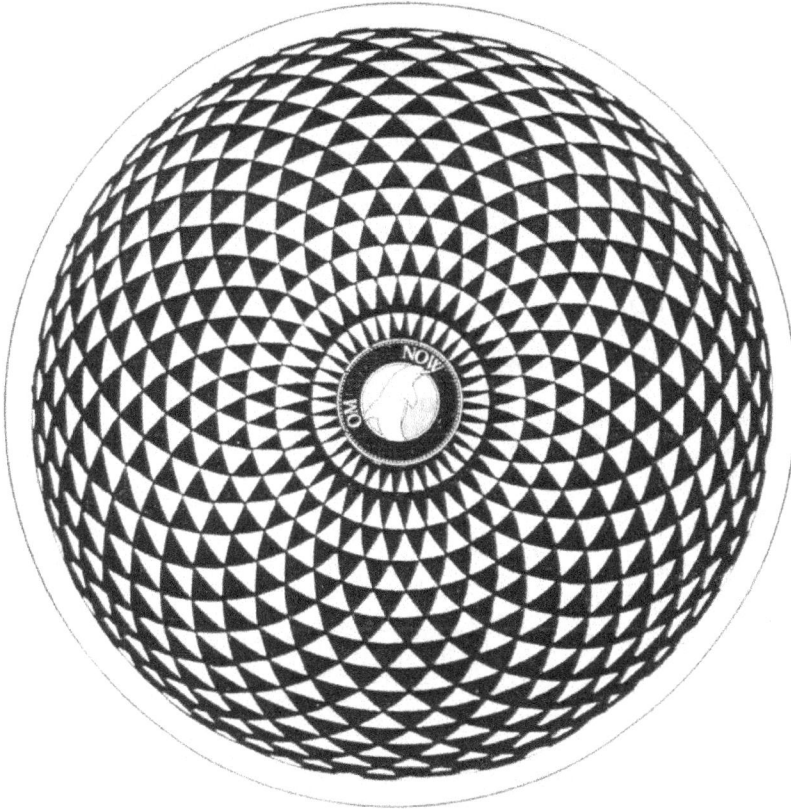

Fig 1b
"Equal Spin" 36 spirals going one way clockwise,
and 36 going the other wayanti-clockwise.
(This is not Nature, this is Equal Spin)
(be careful of Indian Yogis who teach Equal Spin to limit your Crown Chakra. They secretly place this picture on top of your head, and you pay a lot of money for this!)

The true flag for the future One World will be based on Nature's Law, and is shown below in Fig 2, from Part 1, at the very beginning of this article:

Fig 2 is shown again, as I believe that this is the most powerful or psycho-active diagram we have, Even though it is a computer rendition of the counter-rotating whorls, it is unbeatable and perfected as a badge that all young scientists and mystics can wear,
an emblem from the halls of ancient knowledge.
Squint your eyes so you can verify the count of 34 clockwise spirals and 21 counter-rotating spirals.

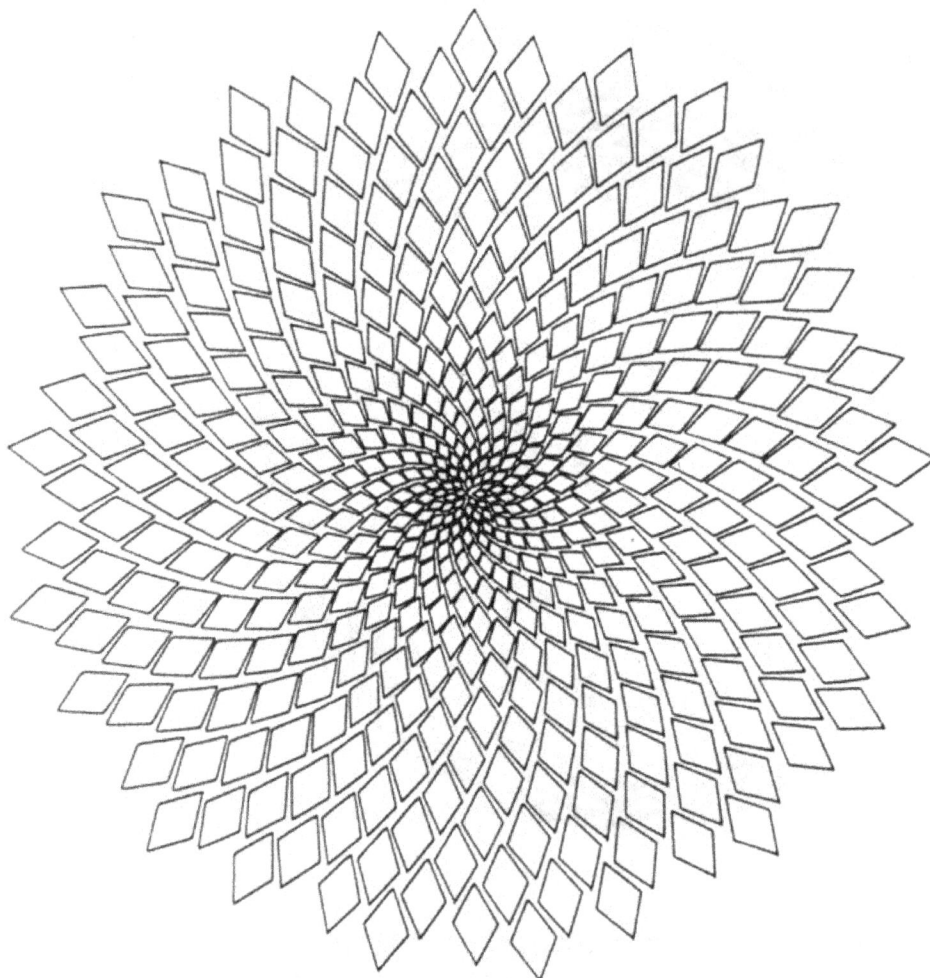

The computer generated 21:34 of the sunflower floret.

PHI FIBONACCI CODE (aka PHIbonacci Sequence)

0 - 1 - 1 - 2 - 3 - 5 - 8 - 13 - 21 - 34 - 55 - 89 - 144 etc

This is where the One becomes the One, not the One becomes the Two, and is expressed by the first two numbers: **"1 - 1"**
One becomes the One, is symbolic of the Creator creating another Self to view the Creation of Self as it expands out in Life.
The **One becomes One**, is what we call **Self-Similarity**, one of the main definitions of "**Fractality**".
The Fibonacci Sequence, as it is incorrectly know, is really a **Trinary** Sequence: where three (**3**) components are considered:

Past + the **Present** = or Creates the **Future**:

$$0 + 1 = \mathbf{1}$$
$$1 + 1 = \mathbf{2}$$
$$1 + 2 = \mathbf{3}$$
$$2 + 3 = \mathbf{5}$$
$$3 + 5 = \mathbf{8}$$

(fractal of 3 Infinite Spirals)

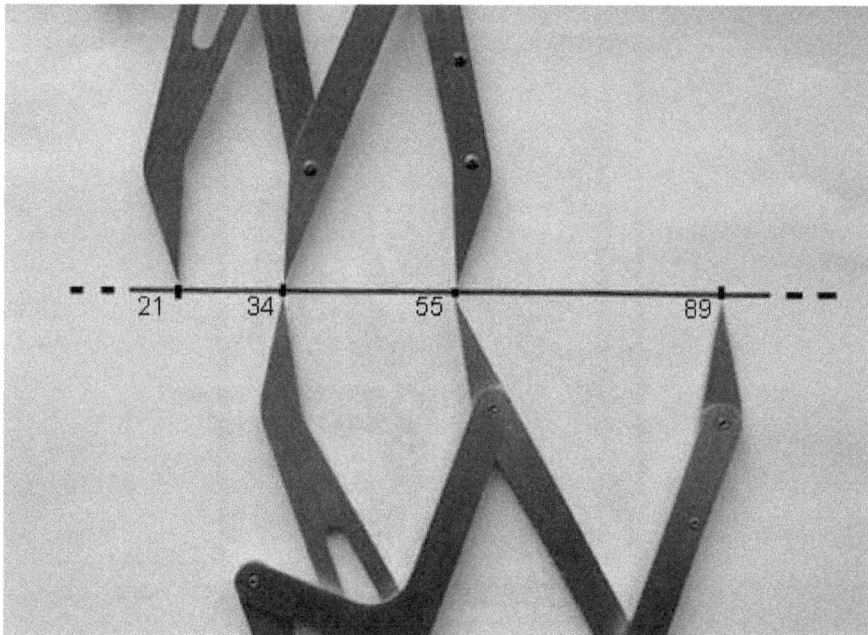

Fig 2a
Fibonacci Calipers
showing the ancient concept of Trinitizaton

PHI RATIO SHOWN GEOMETRICALLY IN THE BINARY CODE:

(taken from: **http://www.phinest.com/**)

Concentric Circle Construction:
Here's a construction using 3 concentric circles whose radii are in a ratio of 1:2:4 which is really a geometric form of the binary code. Refer to Fig 3 on next page.

Method:
Draw a tangent from the small circle through the other two, crossing points A and B and extending to G.
The ratio of the length of segment AG to segment AB is Phi, or
1.6180339887...

Proof:
AB = 2 * $3^{1/2}$ and AG = $15^{1/2} + 3^{1/2}$, which by factoring out the **$3^{1/2}$** can be reduced to a ratio of 2 to ($5^{1/2}$ +1), or Phi.
(nb: the exponent of ½ (or a half or .5) in $3^{1/2}$ means the **Square Root** of 3)

Fig 2c
Christ as Cosmic Architect

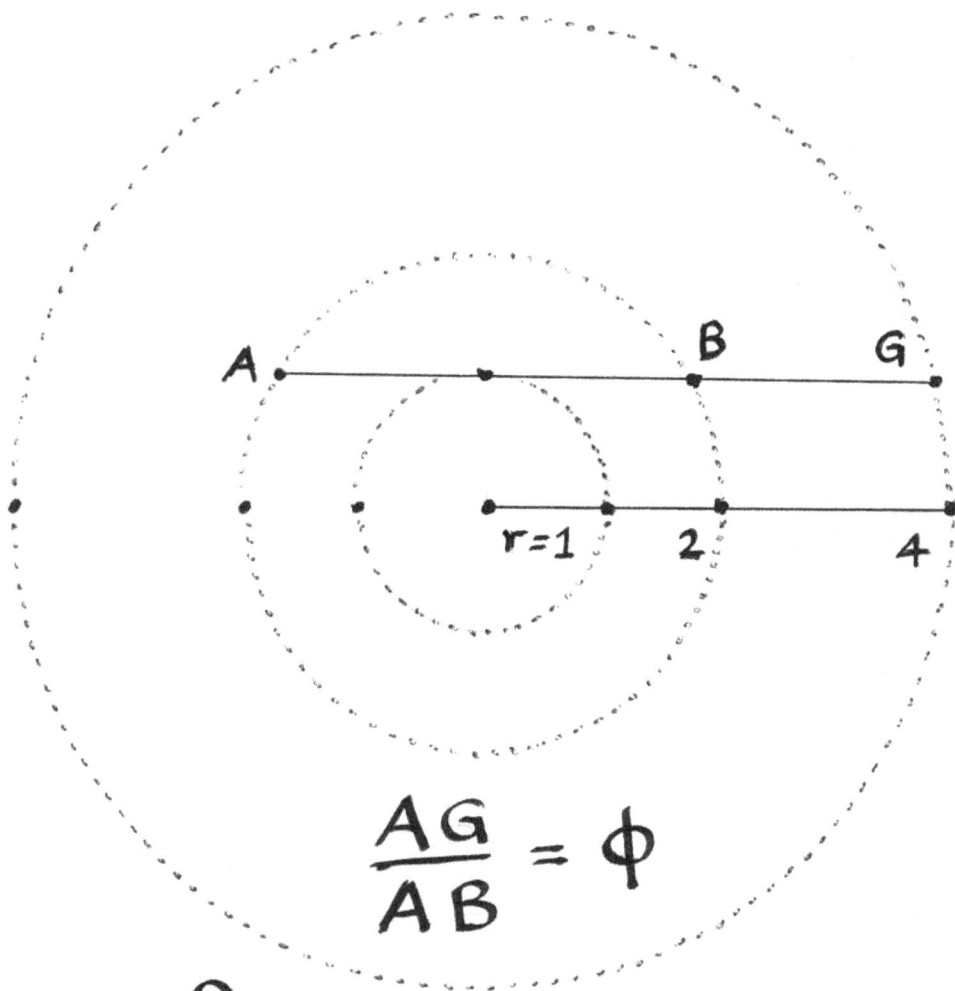

$$\frac{AG}{AB} = \phi$$

Geometric Proof that Phi exists in the Binary Code.

Fig 3
**The Rings of the Binary Code (1:2:4:8)
are indeed in a Phi relationship!**

(this construction was developed by **Sam Kutler** and submitted to www.phinest.com by Steve Lautizar)

This amazing diagram basically proves that the Phi Ratio is in everything, even the dreaded binary code! The Binary Code must now be considered as an Anointed Series or as Galactic Maths.

GEOMETRIC and MATHEMATICAL PROOF for the BINARY / PHI CONNECTION
(utilizing **Pythagorean Theorem** $3^2+4^3=5^2$):

Construct 3 concentric circles that represent the doubling sequence of 1:2:4
Let "O" be the Origin or Centre
With Radius = 1 unit, make a circle
With Radius = 2 unit, make a circle
With Radius = 4 unit, make a circle
Draw a vertical from "O" to touch the Unit Circle at "C"
At this point of C draw a long horizontal tangent that touches the second circle at A and B, top left and top right respectively, and touches the third circle at "G".
OA = OB =2
CB = AC = sq.root($2^2 - 1^2$) = sq.root(3)
AB = double this value = 2sq.root(3)
Find CG
OG = 4
CG = sq.root($4^2 - 1^2$) = sq.root(15)
AG = AC + CG = sq.root(3) + sq.root(15)
Examining the Proportion of the whole length AG to AB
AG / AB = sq.root(3) + sq.root(15) ÷ 2sq.root(3)
Dividing top and bottom by sq.root(3)
= sq.root(3)[1 + sq.root(5)] ÷ sq.root(3)x2
= [1 + sq.root(5)] ÷ 2
= 1.618033988749884
= ϕ

"The Universal Principle of Opposites postulates that everything transforms itself into it's opposite state"

The BINARY CODE in CELLULAR DIVISION

It must be noted here, before we move onto Phi, how efficient the Binary Code truly is. According to an observation by **Drunvalo Melchizedek**, in his now classic book "The Ancient Secret of the Flower of Life" he draws our attention to look at how quickly the first 10 "**Powers of Two**" grow, which is akin to the pro-nucleus' mitotic divisions of cell growth

1 – 2 – 4 – 8 – 16 – 32 – 64 – 128 – 256 – 512 –

You can see that the tenth term is 512, or we could now visualize that there are 512 distinct cells growing.

There are 10^{14} or 100,000,000,000,000 or 10 trillion cells in the human body, and they are constantly being replaced every second. Imagine who long it would take for you just to count this! Yet, the human body has to replace say millions of dead red blood cells every second, which is a phenomenal task. The solution to efficiency is based on the binary code's ability to stack up high numbers in a short time. Have a look at the next ten cell division, and you will be amazed at the staggering half a million of cells created! (Shown below as the number 524,288).

1,024
2,048
4,096
8,192
16,384
32,768
65,536
131,072
262,144
524,288

Now, can you predict how many red blood cells we would have in the next 10 cell divisions after this marker?

Here are the following next 10 binary numbers:

1,048,576
2,097,152
4,194,304
8,388,608
16,777,216
33,554,432
67,108,864
134,217,728
268,435,456
536,870,912

Which is what number? It is now gone beyond half a billion cells?

This is Nature's efficiency at it's best. So we can't bag or bad press the binary code at it too has it's gift, most often known for it's efficient use of number crunching when using the eclectronic calculator, all based on these Powers of 2.

Another fact that is amazing, and taken from Anna C. Pal and Helen Marcus Roberts works in "Genetics, It's Concepts and Implications" is that it takes **46 mitotic cell divisions** to approach the number of cells the human body which is approximated to the 100 trillion count or 10^{14} cells! It's quite curious here that at the 46th cell division, the number of human body cells are equalled and that there are **46 chromosomes in the average cell**.
You can see in the chart below that:
10^{46} = 70,368,744,177,664
which is just under the 100 trillion human cells.

I am going to index all these numbers into a chart soon, in Fig 3, as it is starting to get more interesting as we walk up these laddered rungs of the Powers of 2. So far, the last number is recorded as:
$2^{29} = 536,870,912$

I would like to give you, for reference sake, all the Powers of 2 up to 2^{62} (a huge 19 digit number) and invite you to ponder upon it's association to the Cosmic Vibrations of Touch, Sound, Electromagnetic Radio, Microwave, Heat or Infra-Red wavelengths, all the way down to our visible Light Spectrum, UltraViolet, X-Ray and Gamma or Cosmic Rays... The following chart was found in an old Masonic Library and has carefully been compiled by the AMORC College in association with top physicists and chemists. The chart has on it's back all the musical frequencies relating to the Powers of 2, (with rate of Cosmic Vibrations in cycles per second, and relation to Colours, Chemicals and Vowel Sounds) but I have not given them here. There is a graphic: The International Seal of the Supreme Lodge of the Rosicrucian Order, San Jose, California, showing the "**Rosy Cross**" with the Latin words written circularly:

"Ad Rossam per Crucem".
The Rosicrucian Cross

COSMIC VIBRATIONS

Octave	Manifestation	Vibrations Per Second	Solar Spectrum	Solar Spectrum
1	Touch	2	From 2^{48} to 2^{56}	From 2^{48} to 2^{56}
2		4		
3		8		
4		16		
5	Sound (nb: these boundaries or intervals or numbers are only approximations eg: the lower boundary could be 2^{14} or 2^{15}).	32		
6		64		
7		128		
8		256		
9		512		
10		1024		
11		2048		
12		4096		
13		8192		
14		16384		
15	ELECTROMAGNETIC (Longer Radio Waves)	32768		
16		65536		
17		131072		
18		262144		
19		524288		
20		1048576		
21		2097152		
22		4194304		
23		8388608		
24		16777216		
25		33554432		
26		67108864		
27		134217728		
28		268435456		
29	ELECTROMAGNETIC (Microwaves)	536870912		
30		1073741824		
31		2147483648		
32		4294967296		
33		8589934592		
34		17179869184		
35		34359738368		
36		68719476736		

COSMIC VIBRATIONS

37		1.37439E+11		
38		2.74878E+11		
39	Heat or InfraRed	5.49756E+11		
40		1.09951E+12		
41		2.19902E+12		
42		4.39805E+12		
43		8.79609E+12		
44		1.75922E+13		
45		3.51844E+13		
46		7.03687E+13		
47		1.40737E+14		
48		2.81475E+14		Solar Spectrum
49	Visible Light	5.6295E+14		From 2^{48} to 2^{56}
50		1.1259E+15		Includes
51		2.2518E+15	Spirit Electrons	Infrared / Potassium
52	UltraViolet	4.5036E+15	Spirit Electrons	Deep Red / Lithium
53		9.0072E+15	Spirit Electrons	Orange / Sodium
54		1.80144E+16	Spirit Electrons	Yellow
55		3.60288E+16	Spirit Electrons	Thal / Green
56		7.20576E+16	Spirit Electrons	Stron / Blue
57	X-Rays	1.44115E+17		Violet /Pot
58		2.8823E+17		UltraViolet
59		5.76461E+17		Psychic Blue
60		1.15292E+18	Psychic Projections	Spirit Electrons
61		2.30584E+18	Psychic Projections	
62	Gamma & Cosmic Rays	4.61169E+18	Psychic Projections	
63			Psychic Projections	
64			etc	
65				
66				
67				
68				
69				
70				
71			Psychic Projections	2^{42} to 2^{63} is the
72			Soul Essence	limit of average
73				human Vision
74				

COSMIC VIBRATIONS				
75				2^{40} to 2^{74} is the
76				limit of vision
77	Gamma & Cosmic Rays			Of developed
78				Psychic Sight
79				
80				2^{41} to 2^{73} is the
			Soul Essence	Range of Photo
				Cell Activity

Fig 4

Powers Of Two defining the Electromagnetic Spectrum etc (copied from a rare Freemasonry chart)

JAIN. 29/5/1996   Temptation Ck. "Elisso Sun-Baking on rock".

**Art of Jain 1997
"Laying in the Cosmic Currents"**

BINARY	PHI CODE
Binary Doubling Sequence	Trinary Sequence
1 – 2 – 4 – 8 – 16 – 32 – 64	1 -1 -2 -3 -5 – 8 – 13 – 21
The One Becomes Two	The One Becomes One
As in Human Cell Division	As in The Plan of Plants and Planets, Human Canon
Cosmo-Genesis	All bio-Proteins are Pent
Symbol = 2^n Two to the Power of 'n'	Symbol = Phi = ϕ Adoration of the Pentacle
Limited	Unlimited
Mindless Storage of Data	EnGrailed: is someOne who Understands this Universality of the Fibonacci Sequence.
Energy of the Technological / Machine	Energy of the Etherical / Spiritual
External MerKaBah	Inner MerKaBah
Age of the Computer	Age of the Third Eye Awakening
Mathematics on the Electronic Calculator	Mathematics performed in The Inner Mental Screen
Fear	Love
Control and Separation	One World: Globalization
Fractionation	Fractal
Political Symbol = Pyramid of Mutli-Level-Marketing (MLM) Hierarchy, greed, plutocracy, deception	Political Symbol = Torus allows no Hierarchy! Equality of all Professions, Holographic Unity
Many Fractured Religions	One Universal Religion
Symbol = Equal Spin as on Sahasara Crown Chakra: Turns motor on, but no Travel,	Symbol = Nature's Sunflower Code of Phi-Ratio Spin 21:34 Permit's Time Travel from Macro to Micro
Secret Initiations	No Secrecy
Indoctrination	No Doctrines
Morales	So What
Regulations	All is Possible, Permissible
Factory-Style School Maths	Jain Mathemagics and Vedic

Binary Versus Phi Chart Continued....

BINARY	PHI CODE
Representative Image = Pyramid	Representative Image = Torus
Capitalist	Communist
Non Compressible	Compressible
Explodes	Implodes
Can be Numbered (with Whole Numbers)	Can not be Numbered (Irrational, Infinite Decimals)
Voting via Machines, Lies	Voting via Human Count, Truth
Language of Light	Language of Love
Systems	No System
Electronic School Calculator with Pi button	Electronic School Calculator with true value of Pi and Phi button
Keep adding to this list:	

Fig 5
**Binary Versus Phi in tabulated form
(Or how my quest for the truth of maths
turned me into a communist!).**

Jain's ideal of GLOBALIZATION
or
SPIRITUAL COMMUNISM
One State
One Human Society
One Monetary System
One Language
One Mathematics
One Train Track Gauge

Interestingly:**Binary Versus Phi morphs to Binary=Phi**

The DOUBLING SEQUENCE
And The MATHEMATICAL ORIGIN Of The VW SYMBOL!

Here are some exciting worksheets and notes on the Doubling Sequence, from my book The Magic Of Nine, how it is digitally compressed to form the sequence: **1 - 2 - 4 - 8 - 7- 5** and when plotted on the **9 point circle**, forms the origin of the VW symbol, the People's Car, adopted by **Hitler**, and is the essence of the Binary Code, suggestive of control, manipulation. Whoever owns the essential eternal mathematics, like prime numbers and pentacle harmonics and binary codes owns the world. We are here to liberate ourselves from these ancient controls by renewing our minds and infusing our curriculum with Galactic Mathemagics.

(This touches on another topic:
The Mathematical Origin of Sacred Symbols or
The Demonization Of Anointed Symbols).

Fig 6
**The VW symbol of the People's Car,
is directly based on the mathematical compression
of the Infinitely Repeating Binary Code**

THE MAGIC OF NINE
HIDDEN WITHIN THE DOUBLING SEQUENCE:

2^n Powers of 2	DOUBLING SEQUENCE	ADD DIGITS	REDUCED SUM to a single digit
2^0	1	1	1
2^1	2	2	2
2^2	4	4	4
2^3	8	8	8
2^4	16	1+6 =	7
2^5			
2^6			
2^7			
2^8			
2^9			
2^{10}			
2^{11}			
2^{12}			
2^{13}			
2^{14}			
2^{15}			
2^{16}			
2^{17}			

Fig 7

The Worksheet for the Digital Reduction/Compression
of the Doubling Sequence/Binary Series

* Write down the actual numbers that display Recursion or Repetition:

....

* What is the Periodicity of this Doubling Sequence? ....
(That is, how many numbers are actually repeating, infinitely).

* What happens when you cut this Sequence in half?

1	2	4
8	7	5

* Can you remember all these numbers of this Doubling Sequence?
* Improve your **Memory Power**, and therefore your **Confidence**, by learning this sequence as far as it is comfortable to do so.

ANSWERS:

a) 1, 2, 4, 8, 16, 32, 64, 128, 256, 512, 1024, 2,048, 4,096, 8,192, 16,384, 32,768, 65,536, 131,072

POWERS of 2 2^n	DOUBLING SEQUENCE	ADDING ALL the DIGITS	REDUCED SUM to a SINGLE DIGIT
2^0	1- **1**		= **1**
2^1	2- **2**		= **2**
2^2	3- **4**		= **4**
2^3	4- **8**		= **8**
2^4	5- **16**	= 1+6	= **7**
2^5	6- **32**	= 3+2	= **5**
2^6	7- **64**	= 6+4	= 1
2^7	8- **128**	= 1+2+8	= 2
2^8	9- **256**	= 2+5+6	= 4
2^9	10- **512**	= 5+1+2	= 8
2^{10}	11- **1,024**	= 1+0+2+4	= 7
2^{11}	12- **2,048**	= 2+0+4+8	= 5
2^{12}	13- **4,096**	= 4+0+9+6	= 1
2^{13}	14- **8,192**	= 8+1+9+2	= 2
2^{14}	15- **16,384**	= 1+6+3+8+4	= 4
2^{15}	16- **32,768**	= 3+2+7+6+8	= 8
2^{16}	17- **65,536**	= 6+5+5+3+6	= 7
2^{17}	18- **131,072**	= 1+3+1+0+7+2	= 5

Fig 7a

The Answers to the Worksheet for the Digital Reduction of the Doubling Sequence

c) – By studying the data of Digital Compression on the furthermost column on the right hand side, there appears an obvious repetition or recursion of 6 digits:

1   2   4   8   7   5

d) - This is technically called a **Periodicity of 6**:

e) – There does exist a hidden pattern within this string of 6 recurring reduced single digits.
What do you think would happen if you were to cut this Periodicity of 6 in half and review the mathematical data in terms of two horizontal rows of 3 digits?

1   2   4
8   7   5

Keep looking at these 6 digits until the instantaneous magic of **Pattern Recognition** comes like a flash of lightning into your Consciousness.
That's right, you can see that:

$1 + 8 = $ **9**
$2 + 7 = $ **9**
$4 + 5 = $ **9**

1   2   4
8   7   5
————————————
9   9   9

To improve your Memory Power and Confidence it is a good practice to learn the Doubling Sequence as far as you can manage, at least up to 2^{14}.

PLOTTING the INFINITELY RECURRING PATTERN
of the DOUBLING SEQUENCE
UPON the 9–POINT CIRCLE

Plot the Doubling Sequence:
1   2   4   8   7   5
on the 9 Point Circle:

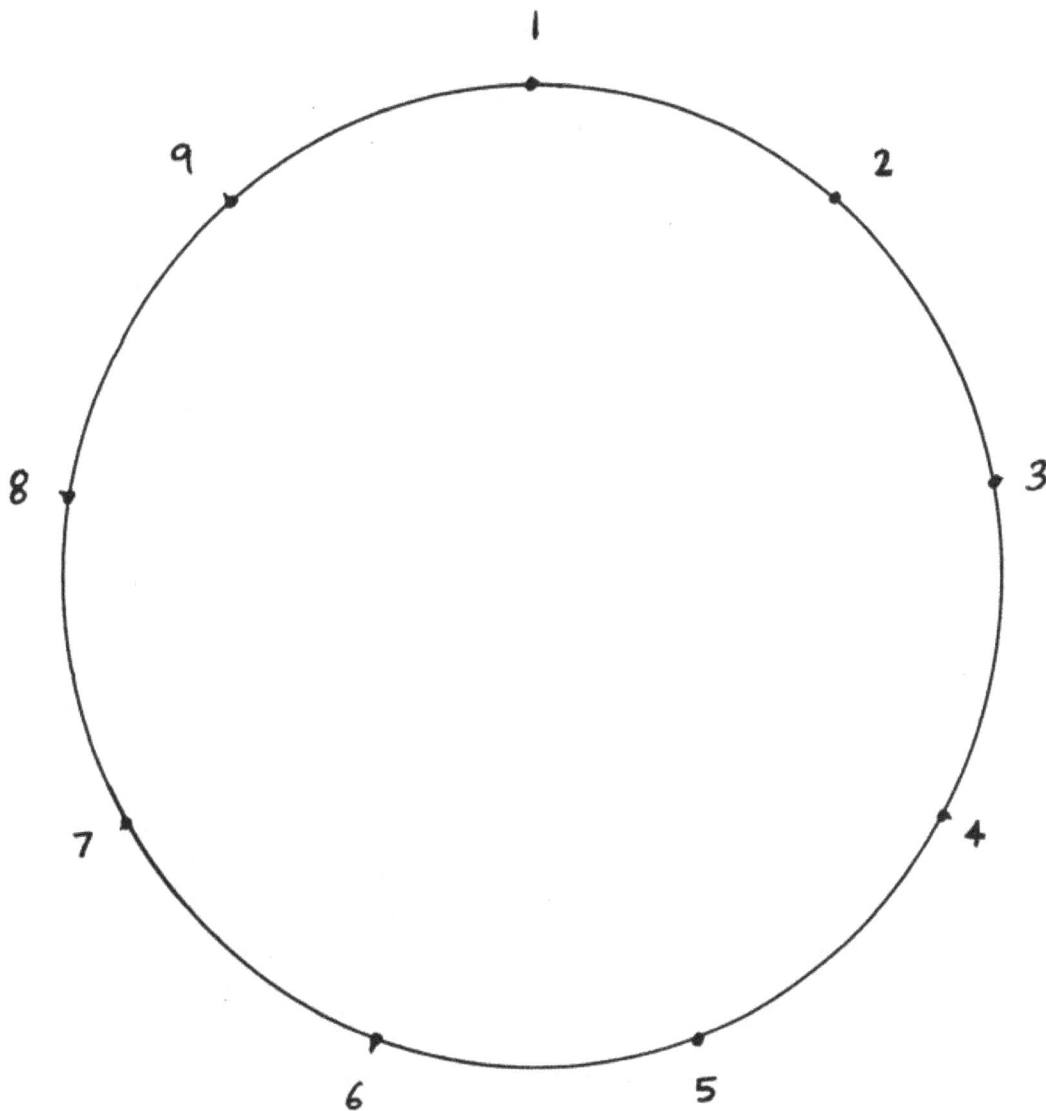

Fig 8
**The Worksheet for Plotting the Infinitely Recurring Pattern
of the Doubling Sequence upon the 9 Point Circle.**

With a coloured pencil or high-light marker, trace over some of the lines just constructed and determine which well-known commercial logo was generated from this design? This subject leads to THE MATHEMATICAL ORIGIN OF SYMBOLS AND COMMERCIAL LOGOS.

➢ Draw a dotted Line in the circle above showing where the **mirror axis** or fold of **symmetry** lies.

➢ What relationship does this pattern have to the 9 Times Table shown in Fig 8 below:

0	9	Draw dotted lines, or mark a dot, to show where the Point of Reflection lies for the 9 Times Table. Are there any other patterns that you can see existing within the 9 Times Table? (Clue: Join similar numbers together, like 1 to 1 and 2 to 2 etc).
1	8	
2	7	
3	6	
4	5	
5	4	
6	3	
7	2	
8	1	
9	0	

Fig 9
Examination of the 9 Times Table

Bahai Temple Lotus Shaped, New Delhi, India, having **9 Petals**.

Answer:

Notice, in the 9 Point Circle of Fig 10 where the dotted Line of Symmetry is, running along from the 9 Point to the midpoint of 4 and 5.

This means that we can fold the pattern upon itself along this dotted line, and therefore displays **mirror-imaging**.

Whatever this means, it is very similar to the mysterious midpoint of the **9 Times Table**, shown below here as Fig 9a, which also lies between the 4 and the 5!

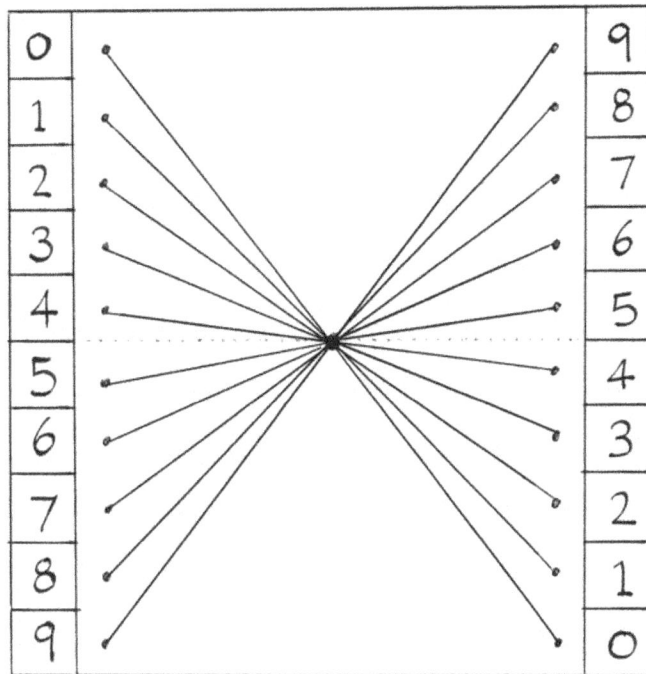

The 9 Times Table (separated into its Pairs (0,9), (1,8) etc & showing the Convergence Point of the midpoint of Pair (4,5).

Fig 9a

Examination of the 9 Times Table, by joining 5 to 5 and 4 to 4 and 3 to 3 etc shows a central node point between the 4 and the 5, similar to the line of mirror-imaging in the 9 Point Circle shown in Fig 10.

Observe the **Mirror Axis** or **Symmetry of Reflection** of the Doubling Sequence:

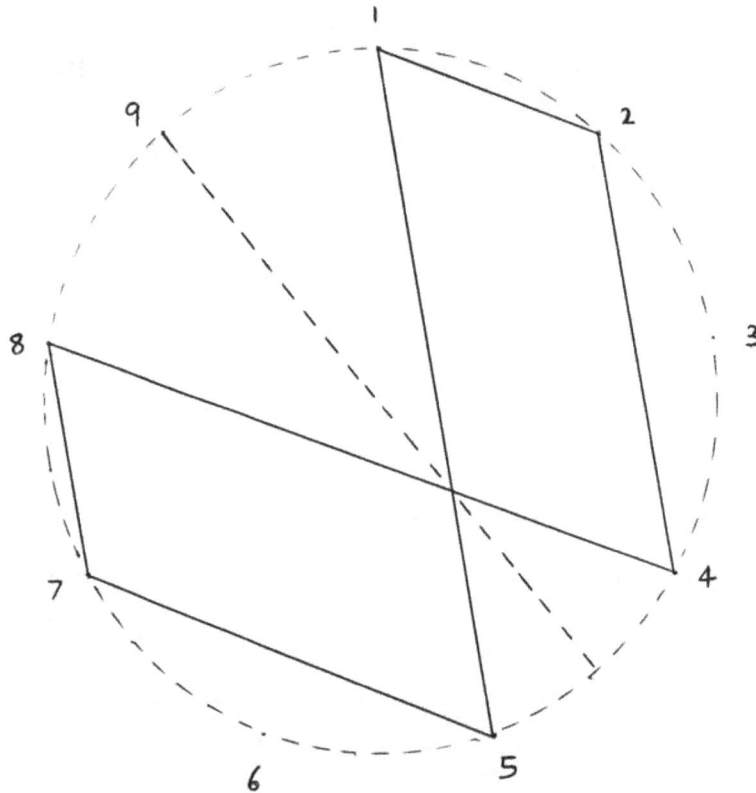

Fig 10
The Answer to the Worksheet for Plotting the Infinitely Recurring Pattern of the Doubling Sequence upon the 9 Point Circle.

The Logo of Mathematical Origin that was consciously derived from this Doubling Sequence in the 9 Point Circle is the **VW** symbol.

Fig 11
The Completed VW symbol derived from the Binary Code

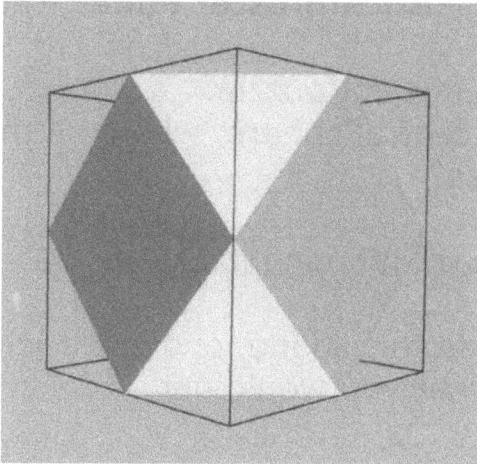

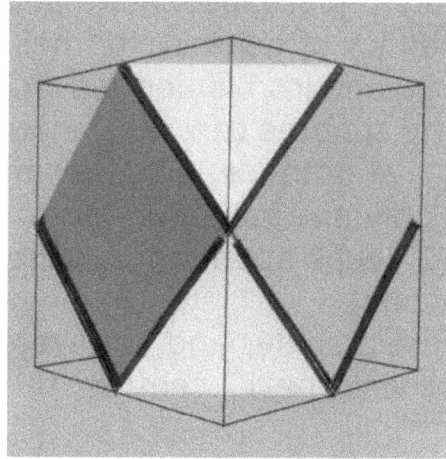

Fig 12a and 12b
The VW symbol (derived from the Binary Code) can be seen in the Hex View or shadow of a 3-Dimensional construct called the CubOctahedron (Vector Equilibrium).

Fig 12a shows the Cuboctahedron and Fig 12b highlights the VW symbol.
(The colour image of the cuboctahedron was taken from the website:
http://www.math.umn.edu/~robertsr
by Prof Joel Roberts, School of Mathematics University of Minnesota, USA)

NOTES ON THE CUBOCTAHEDRON
(Shortened from Cube-Octahedron)
This is a distinguished polyhedron that has 6 square and 8 triangular faces. In sacred geometry jargon, Fig 12a is called the **"hex view of the CubOctahedron"**, which means that when we shine a laser or light upon this 3-D shape, and tilt it around, we can line it up until we see this perfected 6-sided hexagon.
The Cuboctahedron is created by joining the **12 centres** of 12 spheres around a central sphere as shown in Fig 12c:

Fig 12c
a 3-Dimensional construct called the CubOctahedron
(female form, due to spheres or curves)

It is a truly remarkable shape adored by **Buckminster Fuller** who left the legacy of geodesic domes.

The CubOctahedron is derived from an ancient puzzle:

How many spheres can pack around a central sphere, where all the spheres, like oranges, are the same size?

There is only one solution: **12**

(**Base 12** keeps appearing all the time, reinforcing it's status as a galactic code).

If you were a spider and you were inside this shape, and could join with your web the 12 centres, this is what you would get.

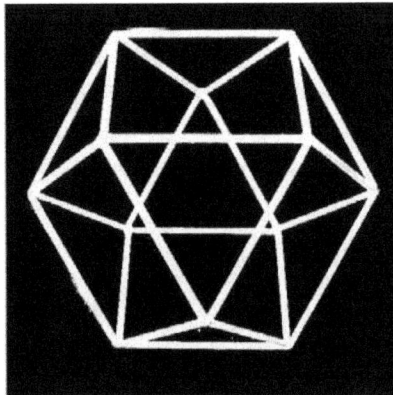

Fig 12d

A 3-Dimensional construct called the CubOctahedron aka Vector Equilibrium, (masculine form due to the stark straight lined geometry). The 12 internal vectors radiating from the centre are all exactly 60 degrees! therefore very balanced. An Honest Shape.

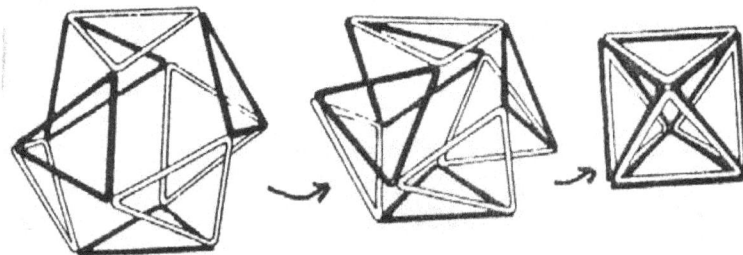

Fig 12e

The Cuboctahedron is compressible, and can form the 5 Platonic Solids. No other shape in the universe can do this!

It's fractal, as "the inside is the same as the outside", since it's internal vectors from the centre to the 12 outer vertices or corners = 1 unit, the same as the 14 outer edges. No other shape in the Universe has the unique property!

It's uniqueness is that when it is compressed or squeezed, when made with dowelling and flexible rubber joins, (as shown in Fig 12e) it forms the **5 Platonic Solids**, thus it was deemed "**shape-shifting**" or "**alchemical**" and referred to as the king or queen of all shapes. Fig 12e shows that with half a twist, the **icosahedral** vertices are suggested, then on full compression the shape morphs to the basic building block of **atomic structure**: the **Tetrahedron**.

PLATE VI

SOLID RHOMBIC DODECA-HEDRON

SOLID REGULAR HEXA-HEDRON PLUS SOLID REGULAR OCTA-HEDRON NORMALLY INTER-LINKED

SOLID REGULAR HEXA-HEDRON (LEFT) AND SOLID REGULAR OCTA-HEDRON (RIGHT)

Fig 12e
The fusion of the cube and octahedron,
shown at the base of the image,
make the Cuboctahedron
(**image taken from** www.gicas.net/poliedri_text.html **by Ugo Adriano Garziotti called: "Polyhedra The Realm of Geometric Beauty").**

Also, when this shape (Fig 12d) is **stellated** (ie: triangular and square pyramidal shapes glued to the faces and making it to grow larger) and you join the **14** new vertices, it creates a beautiful pattern of **12** diamonds called a "**Rhombic Dodecahedron**", and is shown at the top of Fig 12e. Each diamond, has it's shorter axis compared to the longer axis are in the ratio of 1 : the **square root of 2** about 1:1.414... so really it has 12 root 2 Diamonds in this next generation. This is yet another indication of Nature's natural choice for Base 12.

Technically speaking the **Dual of the Cuboctahedron** is the Rhombic Dodecahedron, since each shape forms the other when plotting or joining the face centres.

It can also be formed by stellating the cube with 6 other cubes, as show in Fig 12f and then joining the 6 centres of those cubes.

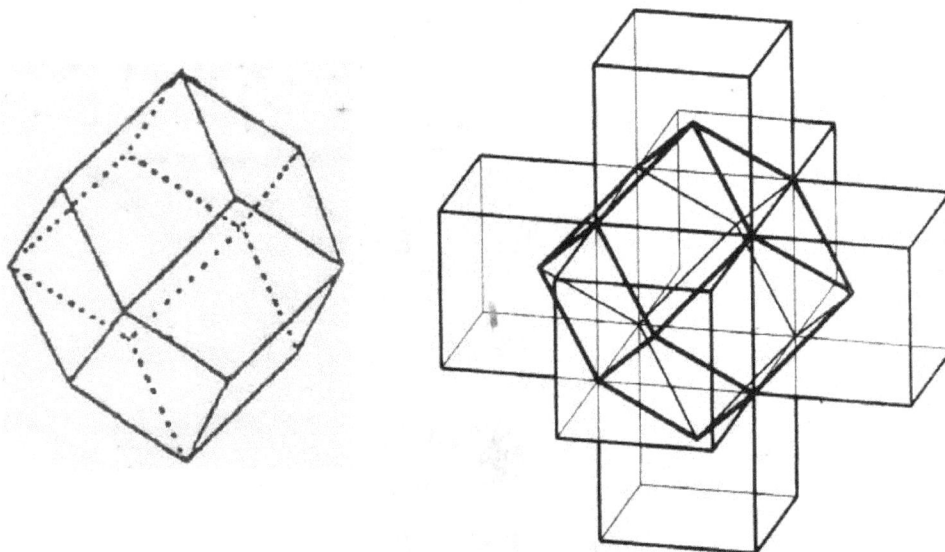

Fig 12f
The Rhombic Dodecahedron having 12 Diamonds or Rhombi formed from the expansion of the cube.
Garnet crystal also shares this rhombic dodecahedral form.
(The Dual of the Cuboctabhedron is the Rhombic Dodecahedron, meaning that this shape is formed when the 14 square and triangular face centres of the Cuboctabhedron are joined).

The reason I am presenting this maths here, is that Phi researchers are looking for Phi in such instances and record such data. It so happened that there was a magazine called the **Fibonacci Quarterly** that had documented the last 40 years of any info on the Golden Mean. One of the visionaries and core members was my colleague physicist **Chris Illert** of Australia. The bad news is that the records from the first 20 years, essentially the most interesting electronic data, has been wiped out of the earth's memory, during the reign of the George Bush government. I know this as a fact as fellow Phi Teachers have approached the **University of Arizona** enquiring about these records, and we have been told directly that it's getting phased out or being watered down. The current and last 20 years of records, as I read the maths on this is appallingly hard to understand, actually I don't understand it, it's written in a new gibberish of mathematics that is sad to teach. This is my attempt to open new and clearer channels into the essential dissemination of the **Phi Records**. The ultimate definition of Phi, if you had studied a whole weekend with me, would conclude on one word: **the ability to SHARE**. Since the old Fibonacci Quarterly is not Shareable, be very cautious about what they are filtering out, and what parts do not get released. That is why you have never seen in print, much of this information that I have garnished here, especially on the **Powers of Phi,** which is essentially the harmonic **Stairway to Heaven** and now known as the **winged Cadduceus** symbol (is discussed in full in The Book Of Phi, volume 5).

The reason why I have introduced to you the shape of the cuboctahedron is that it is connected to the Phi Code 108. The latter Pattern is based on **24** infinitely repeating compressed single digits, so what vessel or container or shape has 24 faces or windows to welcome this data. Here are the 24 equiangular triangles of the 6 Square faces, and the **24** equiangular triangles of the 8 Triangular faces of the sacred **Stellated Cuboctabhedron of 48 equiangular triangles**, looking into it's centre with our x-ray eyes. See Figs 12h and 12i.
We know that the equivangular triangle has 3 lots of 60 degree angles, meaning all 3 angles are the same; so imagine how special a shape can be when it's external vertices are the same as it's internal vertices. When Dan Winter, Phi Scientist, says a definition of fractality is "to make the inside the same as the outside" then here is a geometric association to this phrase:

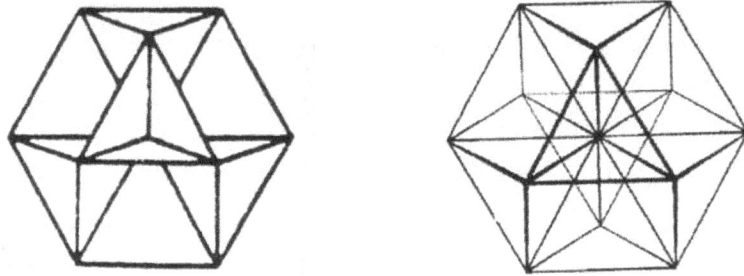

Fig 12g
The 24 edges of the Cuboctahedron,
and therefore establishing it's hidden link
with Phi Codes 1 & 2 which are both based
on distinct 24 Repeating Patterns!
It also can be viewed as a wire-framed model
that possesses 24 equiangular triangles with common
internal and external vertices.

Fig 12h
Stellated Cuboctahedron Jewellery by Jiva and Juliette
Carter of Bali and UK. (known to them as the Heart Star).
http://www.thetemplate.org/GeoShop.html

09/23/2006 07:53 am

Fig 12i

The 48 edges of the Stellated Cuboctahedron, composed of 24 triangular faces of the original 8 triangular faces, and the 24 triangular faces of the original 6 square faces.

The model is being held by Spiros Spiros (or Ross, who studied 8 years with the late Prof Brown of Brown's Gas fame who split the water molecule for hydrogen power and new eco-fuel designs), a true hydro scientist fascinated by this honest shape. The model is made of plastic Geo-Shapes that clip together, allowing Spiros and I to co-create this stellated form from the original 8 triangles and 6 squares of the Cuboctahedron.

cHAPTER 4

HOW To CONSTRUCT The GOLDEN SPIRAL, WORKSHEET

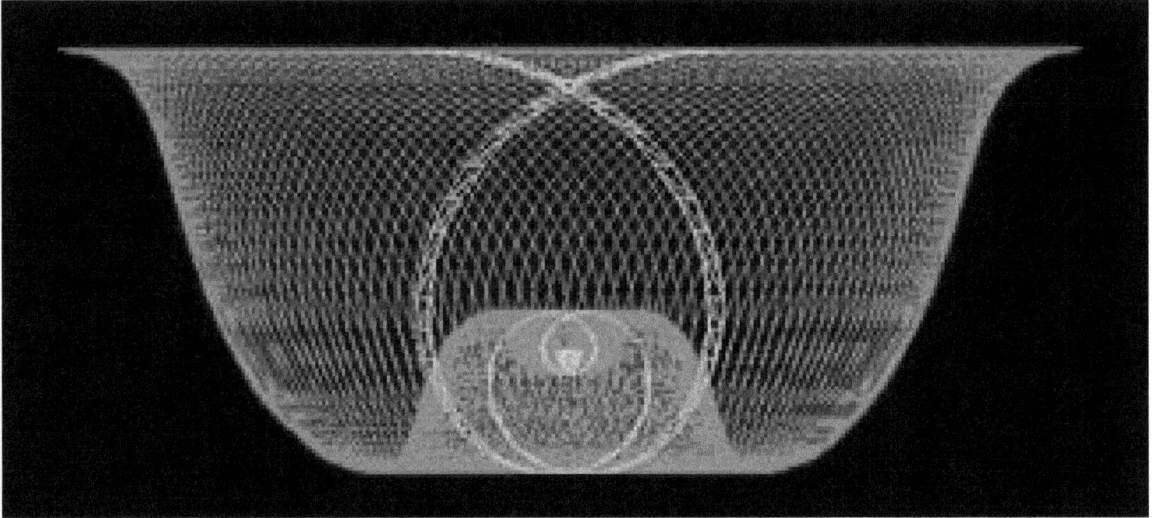

Jain, 06-06-06, Mullumbimby.

(Image: Fractal Heart formed from 2 Phi Spirals:
from the Library of Dan Winter: see hiswebsite   www.goldenmean.info

I, (name)

_____

(State exactly in a few words, or one line, my current and Main
Goal in the Present Tense, that is, that it is already here.)
dedicate this Sacred Geometry today to choose to amplifiy
and magnetize into my Consciousness:

MAIN GOAL: I give thanks that I am/have:

_____

I realize that to achieve this Main Goal, that there are 5 stepping stones that I need to observe. They may be things that I need to add or delete in my life or just be aware of. The best way to list them is to see them in the present tense, focusing only on having this desire in your life now.
I list them below:

Stepping Stone 1:   I give thanks that I am/have:

_____

Stepping Stone 2:   I give thanks that I am/have:

_____

Stepping Stone 3:   I give thanks that I am/have:

_____

Stepping Stone 4:   I give thanks that I am/have:

_____

Stepping Stone 5:   I give thanks that I am/have:

_____

➢ Select an A4 size paper, oriented landscape. Construct AB being 130mm horizontally long. See Fig 1. Let this line be 50mm from the left margin and 50mm above the bottom margin. Our first aim is to make a square on AB. Next is to derive the **Phi Rectangle** from this Square, Next to Subdivide this Phi Rectangle into 5 sub-divisions, Next to construct the Phi Spiral.

All measurements, you will notice, like 210, 130, 80, 50, 30, 20 mm are all in the Fibonacci Sequence multiplied by 10. We are effectively converting the Numbers of the Fibonacci Sequence into a Shape or Art. This creates **Whole Brain Learning** and allows the student to penetrate or go deeper into the mystery of this Sequence based on Nature.

➢ Bisect AB by extending the radius more than half of AB, to say 85mm. Allow these arcs to be placed above and below AB making a fish-like shape. The line between the two intersecting points will determine the **Midpoint** M.

➢ To find the vertical on A, draw a circle around A, letting radius r=40mm.
Extrapolate the line AB to the left edge of the page.
Now open radius of compass to 65mm, make the bisecting arcs above and below the circle. Nb: place your compass point on the actual circle extremities along the original line or plane of AB. Draw in the line through these 2 bisecting points, all the way up to C near the top of the page. C will be the top left corner of the Square, such that AC = 130mm.

➢ Similarly find the vertical on B, by drawing the circle, arcs, giving D the top right corner of the Square, such that BD=130mm.

➢ Using your compass and ruler, check that the square ABCD is 130mm, especially the top horizontal line CD. If this square is not accurate, the whole diagram following will be out.

➢ Before we create the Phi or Golden Rectangle from this square, extrapolate the line AB to the right edge of the page. Softly draw in the Diagonal of half the Square, MD.
Open up your compass radius to MD's length, place compass

point on M and arc down from D to the extrapolated line of AB. E is the point where the arc hit's the extended line of AB.

➢ Extrapolate CD to F such that CF = AE.

➢ Join EF to define the large Phi Rectangle whose dimensions are now 130 x 210 mm.
Did you notice that the Rectangle BEFD is also a Phi Rectangle.

➢ We need to construct 4 smaller squares, all whirling around a hidden, mystically undefinable, though central point "O". This Origin or Center will be revealed later.
To make the next smaller square DFGH at 80 x 80mm, place compass point on F, with r=FD or 80mm, arc this down from D to hit the line EF on G. Similarly, from D, arc down 80mm to hit the line DB making H. Draw Square DFGH.

➢ To make the smaller square 50x50mm, measure GE and use either your compass or ruler to perfect it. This forms EGIJ.

➢ Make an even smaller square 30 x 30mm, forming BJKL.

➢ Make an even smaller square 20 x 20mm, forming HLPQ. It becomes obvious now that the squares are becoming far too small and that inaccuracies will arise. We need soon, on our next sheet, to determine where the mystical center is.

➢ Use this Golden Rectangle of 5 diminishing square subdivisions as a template of 15 dots to impress upon another blank page. Best to use a hard biro point and have several sheets of paper underneath your working page to assist in making easy to see indented points when the template is removed. This will create the template of Fig 2.

➢ Continue to make the 5 diminishing arcs. Remember, at this stage, what were your 5 stepping stones are to achieving your Main Goal. Keep these in mind when you are actually constructing the 5 arcs.
With compass point on B, arc down from A to D. As you draw this, remember what you wrote for Stepping Stone 1, to

amplify this intent.

> With compass point on H, arc down from D to G. As you draw this, remember what you wrote for Stepping Stone 2, to amplify this intent.

> With compass point on I, arc down from G to J. As you draw this, remember what you wrote for Stepping Stone 3, to amplify this intent.

> With compass point on K, arc down from J to L. As you draw this, remember what you wrote for Stepping Stone 4, to amplify this intent.

> With compass point on P, arc down from L to Q. As you draw this, remember what you wrote for Stepping Stone 5, to amplify this intent.

> These arcs will continually diminish forever, right down into the atomic world. We need to determine where the Origin or Center of the **Toroidal Wormhole** is. The ancients knew that 2 critically intersecting lines at **90°** to one another determines this Center. See Fig 3.
To locate this Center Point "O", draw two diagonals of the 2 largest Golden Rectangles. Draw CE (of the largest Phi Rectangle AEFC) and draw BF of the next smaller Phi Rectangle BEFD. Notice how they intersect **perpendicular** to one another. The diagonal line CE is the critical angle **32°**. (This is what takes the 3-Dimensional **Cube**, when tilted 5 times at 32° forms the phi-ratioed **Dodecahedron** or 4th Dimensional Cube, Pythagorean symbol of the Ethers or **Spirit**. In this sense, this angle permit's multi-dimensional travel. **The Dodecahedron is the 3-Dimensional form of the 2-D Pentacle**.

> Now that you have determined the Center or Eye of the Spiral "O", you will not need the compass anymore and your aim is to draw freehand the continuation of the spiral of 5 diminishing arcs, so that you seem to tumble around this center as the line gets finer and closer to the mystical center. Must keep in mind your Main Goal. When you have come to a

still point, hold your pen or pencil on that point, using your index finger to hold the top of the pen, and just connect with the geometry and the focused thought you are holding. Be still and feel this center.

➤ All the while we are focusing on our Main Goal.
Our conclusion is to share an open eyed group meditation, watching a dvd: "Visual Symphonies" by **Jonathan Quintin**, depicting the amazing moving graphics of the Torus or Imploded Sphere which is the 4th Dimensional form of the Phi Spiral and part of our **Ascension Process**.

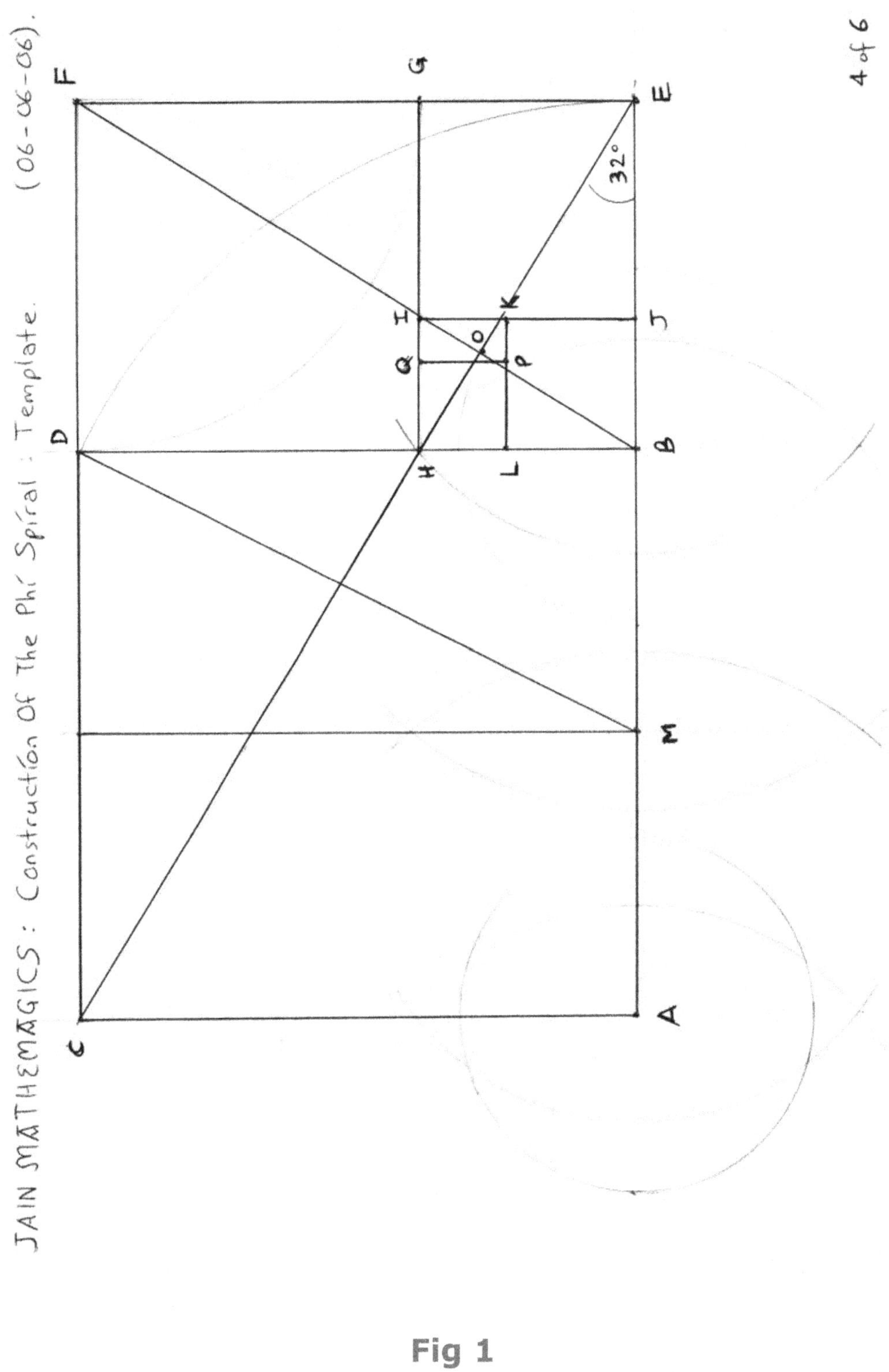

JAIN MATHEMAGICS : Construction Of The Phi Spiral : Template. (06-06-06).

Fig 1

Worksheet 1: **Construction of the Phi Spiral Template,
Beginning with the Phi Rectangle.**

JAIN MATHEMAGICS : Template Of 14 Dots for Phi Spiral Construction.

5 of 6

Fig 2
Worksheet 2: **Template of 14 Dots
for the Construction of the Phi Spiral.**

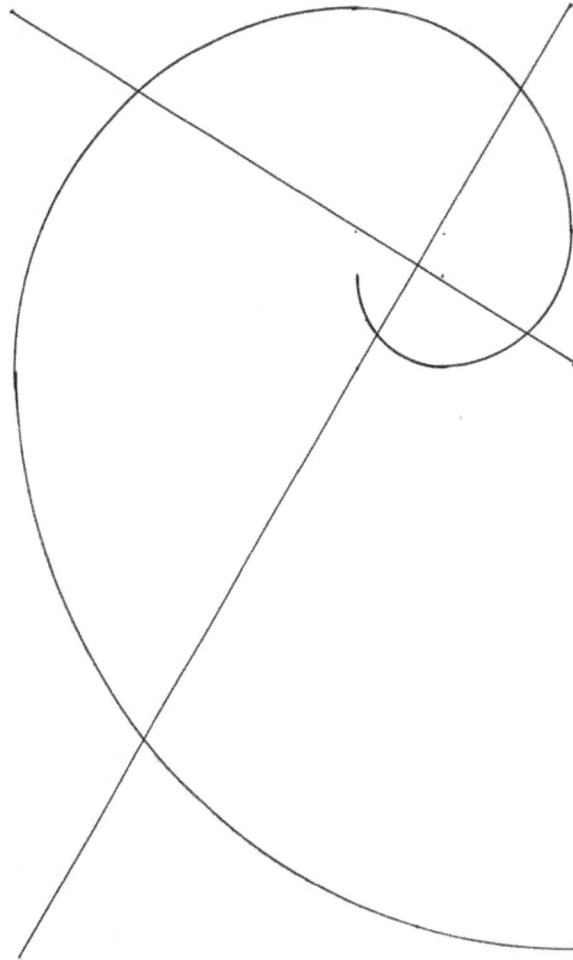

Jain Mathemagics : The 5 Diminishing Arcs Approaching The Mystical Centre. 6 of 6.

Fig 3
Worksheet 3: **The 5 Diminishing Arcs Approaching the
Mystical Centre, for the Construction of the Phi Spiral.
Nb: the whirling travels about perpendicular axes which
permit's the long wave of the universe to harmonize with
the short wave of the atomic reality. It represents the 90
degrees Phase Shift of Consciousness, the ability to enter
or exit from one galaxy and into another.**

(Art of Jain, 1994 "Praying For The Celestial Transcripts")

cHAPTER 5

NUMEROLOGY AS A SCIENCE
JAIN'S LAW OF DIGITAL COMPRESSION
BASE TEN HOMEOPATHICS

By JAIN
(14TH November 2007, Watsons Bay, Sydney)

Numerology As A Science:

Includes the following topics:

⊙ Jain's **Law of Digital Compression**

⊙ The Digital Compression of the Common **4 Times Multiplication Table**.

⊙ The Formation of an **Enneagram** 9 Pointed Star by translating the Periodicity of 9 of the 4X Table into the 9 Point Circle.

⊙ **Converting** 21 **into Binary**
and converting $_{Base\ 10}$ 1000 to = $_{Base\ 2}$ 1111101000

⊙ **Bruce Cathie's Harmonics of Light** based on the **Reciprocal of 144**.

⊙ The **Heart Fractal** from the library of Dan Winter

⊙ **Base Ten Homeopathics**

The study or science of Numbers is an ancient craft that opens windows into the understanding of our universes and consciousness. It shows inherent or hidden patterns underlying number tables or matrixes and sequences, making the invisible visible.

Certain number sequences or tables can be reduced to single digits, like the common 4 times table, the whole 9x9 times table, the Fibonacci Sequence in particular shows exceptional repetition or recursion of single digits totaling a sum of **108**, 108, 108, like a secret pulse to the Vedic universe. There are many other codes which have been dealt with in my other self-published books.

To understand how Numerology is the key to converting Number into Art, let us examine the
**NUMEROLOGICAL COMPRESSION
OF THE COMMON 4 TIMES TABLE**

Let us look into or inspect the hidden or inherent symmetry underlying the well known Four Times Tables.
We will firstly write out the series up to 72, and then numerologically reduce all two-digit numbers to a single digit from 1 to 9. This is achieved by adding the two digits, as in the first case of 12 which equals 1+2 = 3 and
28 = 2+8 = 10 = 1+0 = 1.

4	8	12	16	20	24	28	32	36	40	44	48	52	56	60	64	68	72
4	8	3	7	2	6	1	5	9	4	8	3	7	2	6	1	5	9

Fig 1
**The Digital Compression
of the Common 4 Times Multiplication Table**

The top line of Fig 1 shows the uncompressed numbers of the 4 Times Sequence.
The lower line is the digital compression of same.
Can you see in the lower line that there is some recursion going on, in fact every nine numbers in sequence repeat.
We say that this has a **Periodicity of 9**.
Here is this necklace of 9 infinitely repeating compressed digits:

4	8	3	7	2	6	1	5	9

Fig 2
The Periodicity of 9
in the Digitally Reduced 4 Times Multiplication Table

Then, we will plug this code into the 9 POINT CIRCLE shown previously as Fig 7.

By joining a single, unbroken line from 4 to 8 to 3 to 7 to 2 to 6 to 1 to 5 to 9 and closing the circuit by joining a line from the beginning to the end (**The Alpha and The Omega**) means we are really exploring the meaning of **Order Amidst The Chaos**.

Notice the beautiful pattern that is formed:

the Nonagram or **Enneagram Star**:

Fig 3
The Formation of an Enneagram 9 Pointed Star
by translating the Periodicity of 9 of the 4X Table
into the 9 Point Circle

Some scientists, like Dan Winter,
(see my internet article in my Newsletter Archives
http://www.jainmathemagics.com/newsletter_lookup.asp?newsletter_id=25)
ridicule the Phi Code of 108 as they find it difficult to understand simply the Laws of Mathematical Reduction like 21 = 3 + 1 = 3. They call it bogus and unscientific, non-mathematical. How can 10 x 10 = 100 = 1 + 0 + 0 = 1. They fail to see the deeper meaning of **Unity Consciousness** in the sense that 1 = 10 = 100 = 1,000 = 10,000 = 100,000 = 1,000,000 etc.

The brilliance of modern mathematics is evident in the brilliance of the Binary Code where giant numbers are digitally compressed down to two simple and basic expressions that we call "**Zero and One**" or "**0** and **1**". This is unarguably pure numerology, where the powers of 2, the well known doubling sequence of

1, 2, 4, 8, 16, 32, 64, 128, 256, 512 etc are referred to as a Base 2 system where these higher numbers are reduced to a choice of two single digits, like an on/off switch...

Let us take any multi-digit number like 21 and watch how this higher number is reduced to a series of single digits:

2^4	2^3	2^2	2^1	2^0
16		4		1
1	0	1	0	1
		Fig 4		

The Base 10 number 21 ($_{Base\,10}$ 21)
as a Binary Base 2 Number 10101 ($_{Base\,2}$ 10101)

Notice how the powers of two (the numbers 16, 4 and 1) have been replaced by the number 1, and the remaining numbers (2^3 and 2^1) were not involved in the parts summing to 21, so they were replaced with a zero or non-entity.

We therefore say that 21, (in Base 10) is equivalent to 10101 (in Base 2).

Is not this conversion or transduction from a higher base 10 to the simplest base of 2, a form of **digital compression**, that this higher number 21 is rearranged atomically to 10101 !!!???

This simple Law of Mathematical Reduction that rewrites $_{Base10}$ 21 as $_{Base2}$ 10101 is also a Universal Law of Numerical and therefore Vibrational Compression.

We have effectively reduced 21 whole entities or unit's or things into a compressed number based only on Zeroes and Ones!

This Computer Age that we have been born into, is pure Numberology.

This **Digital Age** is pure Compression.

For another practice, can you convert $_{base\,10}$ **1000** into a binary number sequence?

The most simplest and therefore the most compressed Number Base System is Base 2, Binary.

As just explained, the Binary Code numbers are written in a positional system that uses only 2 digits: Zero "0" and One "1". The only problem with Base 2 is that when it represents the larger numbers, like say the number 1,000 it would need lots of zeroes and ones to express it.

Whereas in Base 10, as you can see here, it takes only 4 digits to express $_{Base\ 10}$ 1,000 and therefore we can say in this example, that Base 10 is actually more compressed and effective than Base 2!

1000 in base 2, the long way =

2^0	= 1	= No	= 0
2^1	= 2	= No	= 0
2^2	= 4	= No	= 0
2^3	= 8	= Yes	= 1
2^4	= 16	= No	= 0
2^5	= 32	= Yes	= 1
2^6	= 64	= Yes	= 1
2^7	= 128	= Yes	= 1
2^8	= 256	= Yes	= 1
2^9	= 512	= Yes	= 1
2^{10}	= 1,024		

2^9	2^8	2^7	2^6	2^5	2^4	2^3	2^2	2^1	2^0
512	256	128	64	32	16	8	4	2	1
1	1	1	1	1	0	1	0	0	0

Fig 5

$_{Base\ 10}$ **1000** = $_{Base\ 2}$ **1111101000**

Sometimes for typesetting reasons, the exponent or index or power is represented by this symbol "^" which means

"to the power of"

thus 2^10 means 2 to the power of 10

or 2^{10} = 2x2x2x2x2x2x2x2x2x2

and since it = 1,024 being greater than 1,000 it is not used to compute 1,000 in base 2.

The Column of Zeroes and Ones on the Right, tells us Yes or No, whether those powers are used. That is, we can see that

1,000 = 512 + 256 + 128 + 64 + 32 + 8
so we mark each of these powers as a "1" or a "Yes"
and all the other numbers or powers listed are therefore switched off "No" and expressed as a Zero "0".
Reading from the bottom to the top:
Thus 1,000 = **1 1 1 1 1 0 1 0 0 0**
Look at the comparison,
In Base 10, we needed only 4 digits to express 1,000
In Base 2, we needed 10 digits to express 1,000
So it is raises the question of **which base is really the most compressed or effective**, or energy-saving.

A great and humble mathematician and free thinking scientist, **Bruce Cathie** has developed new topics based on the Harmonics of Light, Mass and Gravity. It is very interesting, how he takes whole numbers, like the number **144** and examines it's **reciprocal**, which means we divide 1 by 144 which effectively takes the macro-cosmic whole number and reduces it to it's micro-cosmic counterpart or fraction expressed commonly as 1/144 or **1 ÷ 144**.
When we decimalize this fraction it appears as:
.0069444444444... The three dots at the end represents that it keeps going onto infinity. But Bruce Cathie pioneered this scientific numerology to a fine art, by compressing the whole decimal to a distinct whole number by deleting the initial decimal, by deleting any zeroes, and by rounding off the last repetitive number to arrive at **695** which he distinguishes as the **Harmonic of Light**. Bruce Cathie is mainly known as the New Zealand pilot who architected the **World-Wide Grid** of ley-lines encircling the planet by joining the dots of ufo observations. He states clearly the supremacy of our Base 10 system is based on the 9 standing digits 1-2-3-4-5-6-7-8-9 and one Zero which acts as a bridge or connector to the higher powers of 10.
The **Zero** has multi-dimensional power, not only does it demarcate the place value system of unit's, tens, hundreds, and thousands, but it can magnify any number like 4 which becomes 40 which becomes 400 which becomes 4,000 etc
The **Place Value System** is also another form of digital compression, in the sense that when we write the number 4 we are really writing the number 4,000 as show in the example below:

$$4096 = 10^3 \quad 10^2 \quad 10^1 \quad 10^0$$
$$= 4 \qquad 0 \qquad 9 \qquad 6$$

There is an interesting short-cut that quickly compresses the number 4,096 to a single digit, and that is known as "**The Casting Off Of Nines**". In the number 4,0<u>9</u>6 just ignore or delete the number 9, and what is left is 4+6 = 10 = 1+0 =1.

In terms of Vibration or Memory, this number 4,096 is tainted by the additive memory of 4+9+0+6=1. I call this **Base 10 Digital Homeopathy**, in the sense that a number like 4,096 can be reduced to a single digit. To understand this connection between number crunching and homeopathics I will need to go into a deeper discussion here defining what is Homeopathy...

Now, as most of you know, many scientists and doctors scorn Homeopathy. How can a homeopathic remedy be a real science? A homeopathic remedy is taking something like a few drops of **Arsenic**, in a jar of water, shaking it by percussing it (generally done **108 times**!) on the palm of your hand, and tipping out the whole of the contents. What remains on the inside of the jar is just a taint, a memory, not even a drop of the diluted arsenic, yet when the process is repeated again and again, by filling that same jar with water, it is diluted even more and more. Lets call the original arsenic "1", and the first dilution, call it "10X", and the next dilution, by tipping out the contents and refilling with water, as "100X"; and the next dilution by tipping out the contents and refilling, as "1000X". This is repeated and repeated. At the point of stopping, we shall call the substance remaining as our "**Mother Tincture**". To make your own remedy, the patient takes only a few drops of the Mother Tincture to make their medicine.

Now the oddity that arises, in the Law of Homeopathics, is that the more we dilute and dilute, up to and beyond a 1,000,000X the more powerful the vibration, and the more effective the healing. How can this Law of **Vibrational Medicine** be a real science? It is a fact no doubt, this ancient system of healing actually works! Who cares if it is unscientific. It is a **Law of Vibration**, and this Homeopathic dose of poisonous 1,000,000X Arsenic actually heals. The ancient Greek healers called this a **Law of Likeness** or Sameness, that "**Like Cures Like**". It is a bit like the fractal zooming, as we zoom into the centre of a heart, the same **self-similar** shape keeps appearing, **scale invariantly** which means that the size does not matter, whether it is the size of the universe or the atom, what matters is that it is self-similar!

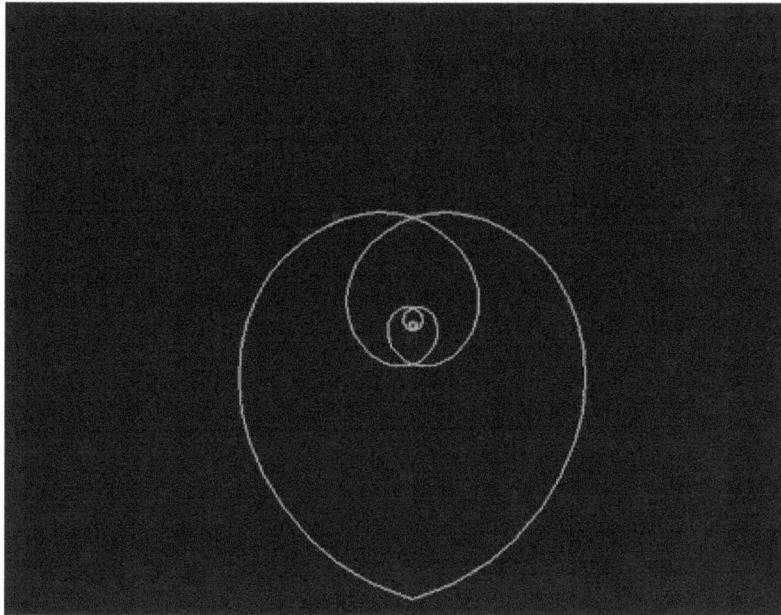

Fig 4
The Heart Fractal from the library of Dan Winter.

(In the actual animation, the more you zoom into the moving centre, the more you keep seeing the same self-similar heart shape. To see the animation type in:
http://www.goldenmean.info
But do not use google to search this as it is owned by the **CIA** who have blocked or witchhunted his website).

In the world of Homeopathics, it means that if you are bitten by a snake, you put the highly diluted self-similar form of the poison back into your body, and the body, that is highly intelligent, knows this Memory or Akashic/Vibrational Record, or recognizes this Geometry and actually makes an anti-dote of the poison to neutralize it. The science of this process is another debate, but the principle is the same in what I call The Law of Digital Compression, that this 1,000,000X Arsenic can be compressed back to 1+0+0+0+0+0+0. No matter how much we dilute it, it is still arsenic.

Numerology is thus a Law Of Vibration, so if your local scientist or Mathematician don't grok (understand) it, just ignore their critical remarks, just dilute them, as it shows either ignorance or immaturity of the facts. It is not expected that a modern scholar understands this lost Atlantean concept that 34 = 3+4 = 7. How ridiculous that 34 = 7 it must be dismissed!

So we can summarize that Numerology is a Higher **Law of Vibrational Numbers**. We could call it "Base Ten Mathematics" in the sense that our bodies were designed with 10 fingers and 10 toes (except for Fred who has 2 digits missing from a lab mishap, and Saraswati of Kolkata who was born with 12 fingers and 12 toes) is really Earth Mathematics, and that the secret (or will I dare to call it a "science") of Compressing Numbers is the Heavenly Mathematics. Or better still:

Base 10 = Earth Mathematics

Numerology = Digital Compression = Galactic Mathematics.

I believe also that this statement to declare Numerology as an Ancient Science of Numbers was so informative about the Laws of our Universe that it too, like the Pentagram, the Swaztika, and the Phi Spiral (3-Dimensionally it is the Ram's Horn) were **demonized** to create **Fear**, not Love, and therefore deny access to the **Memory** of who You truly are!

I have written this as an introduction to the Numerical Nomad or Neophyte, and the Mathematical Monks, to adore and embrace Numerology as the ultimate level of inter-connecting mathematics or **Sufi** Mathematics (universal order that embrace all faiths), that is all-encompassing like the old Indian Jaina Mathematics.

When you believe and grok this Digital Compression, a new window in your consciousness will indeed open up. Digital Compression will make the boring Times/Multiplication Table of 9x9 open up into a vista of **Atomic Art** and elegant eternal symmetries. A mathematician is not a true mathematician if he/she does not appreciate the Digital Compression of this 9x9 matrix that when translated into Art reveals the underlying crystalline nature of our very being.

It is important to remember that one of the ancient 16 Sutras known in total as **Vedic Mathematics**, a system of mental, one-line calculation, is called "Digital Sums" or "Digit Sums" or in our modern context, I call it "Digital Compression". The pope of India, **Bharati Krishna Tirthaji**, before he died in 1960, attributed these 16 sutras as having the ability to solve all know mathematical problems, if not mentally, these sutras could solve mathematical problems via one-line calculations.

By embracing the Digital Compression as one of the spokes of the wheel of True Mathematics, you will appreciate that the Living Mathematics Of Nature is indeed represented by the Fibonacci Numbers: 0, 1, 1, 2, 3, 5, 8, 13, 21, 34, etc that when compressed reveal the two distinct "Shri 108 Codes". Now you have been told for thousands of years that there is no pattern in this Phi Ratio (being the decimalization of one Fibonacci number divided by it's preceding number as in 34 divided by 21, to give the Divine Proportion 1:1.618033...) yet when we digitally compress the **Powers of Phi** (1.618033... to the 2nd, 3rd, 4th, 5th, and nth Powers) another distinct 108 Code is revealed!

So be cautious when your great metaphysical phi scientist educates the masses that:

1 – there is no pattern or mathematical recursion in the phi ratio, (which is contradictory to the very meaning of the phi ratio which is about embedded, self-similar, recursive nestings.

2 – that it is mere senseless numerology.

You can be sure that underlying such negative statements reveal only 2 things:

1 – ignorance of the Higher Laws of Mathematical Reduction

2 – that they are jealous, that as distinguished teachers or metaphysicians, they failed to see or locate the hidden 108 repetitive codes.

I write this article merely to elucidate to my students that Numberology (or Numerology) is a Sacred Science and to ignore the wafflings or derision of any other teachers who demonize it, trivialize it and trash it.

At least any controversy is a good opportunity for you, the reader, the student, the neophyte to determine for yourself, that any Mathematical Truth is self-evident and easy to understand, and above all, Timeless, **Fixed Design** that is **Shareable**.

Regards,   **Jain 108**.

www.jainmathemagics.com
jain@jainmathemagics.com
www.mathemagicsasia.com
www.jain108.com

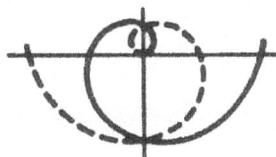

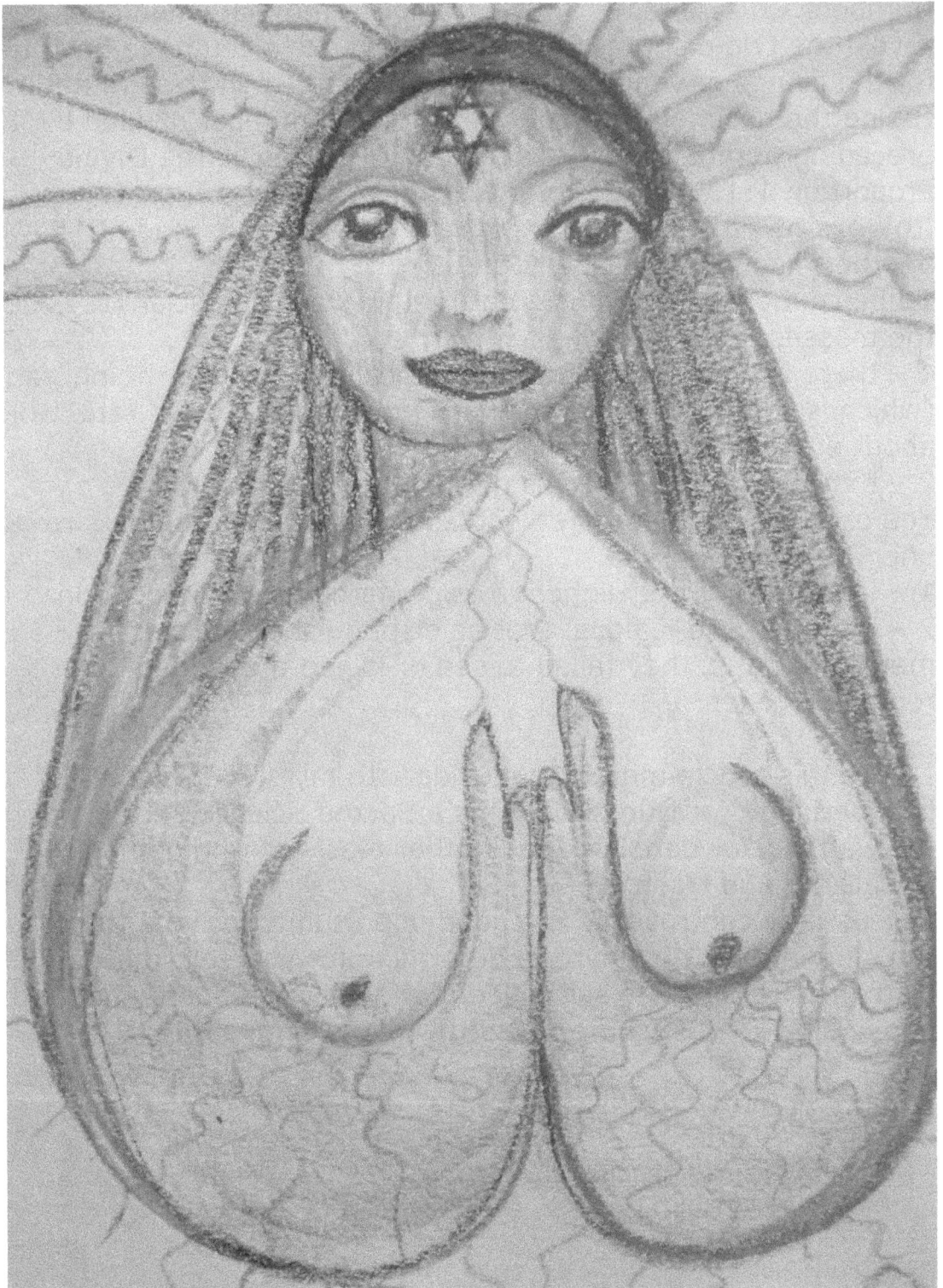

(Art of Jain, 1994 "Angelic Guardian of the 108 Codes")

cHAPTER 6

JAIN 108 Vs PETER SAAD VS HANLEY VS SEPHIRA INCIDENT

2am   15/5/2007

RE: RUBBISHING VEDIC MATHS.
RE: LEARNING TO LOVE

I had just finished a 2 weeks tour lecturing and promoting Jain 108 Mathemagics in Sydney. With my white camry station-wagon all packed with books and paraphernalia, having stayed at Watsons Bay for 2 weeks, I decided to visit my Mother and give her some belated Mother's Day flowers. I had vowed not to stay again at my parent's house in Kirrawee as often, if not always, my father keeps pounding psychologically on me that I am a failure and have wasted my life. It so happened that I did decide to stay overnite just to be with my folks, and I would leave in the morning back to Mullumbimby to unload and then to Murwillumbah to see my Love Sephira 7.5 months pregnant...

... had dinner, shower, phone calls, Mum invited me upstairs to come up for a cup of tea and cake. They were watching the Arabic news. The phone rang, it was for me (regarding our tomorrow rendezvous at Moore Park to view the Leonardo da Vinci machines, with Neena Scott, Hugh Moro (possible sponsor "Lord" in the Magic Square business structure, Mercury and Vedant...). As I was chatting to James Mackney on the phone, I noticed on TV a quick ad saying to stay tuned for a thing about "Math Solution". Straight away I knew it was going to about "Vedic Mathematics" or Rapid, Mental Calculation... I bid James farewell, hung up, and eagerly informed my parents that if they would like to know or get an understanding of what I have done for the last 30 tears, and what I was just teaching in Sydney, that they could stay tuned for the next program that will demonstrate Mental Calculation.

And so it happened, I was correct, and alerted my parents to the fact that the Victorian Professor of Mathematics being interviewed was an imposter, as he did not acknowledge the Vedas, nor the Indian master Bharati Krsna Tirthaji (1884-1960), nor Vedic Mathematics. Besides this, all was well presented, over the next 5-10 minutes, focusing on how young teenage students in Melbourne could do their times table eg:

7 x 8

–3   –2 with circles around both the minus three (-3) and the minus two (-2).

Cross-Subtracting:

7-2 = 8-3 = **5** in both diagonals,
                giving 5 lots of the base 10, giving 50

Multiplying the two deficiencies:

–3 x –2 = 6   giving **6**

adding or combining the two parts = **5 / 6**

thus forming the final answer of 56.

Then they gave another example in Base 100

97 x 96

–03 x –04

= 97 – 4 / 3 x 4
= 93 / 12
= 9,312

The teacher was impressed and the students found it easy.

The program ended with **Channel 9** discussing this with a parent who was impressed by this system that could make their child into a mathematical genius.

After this ended, Dad (Peter Saad) and Mum (Yvonne Khoury) and I commented on what we thought. As expected, Dad was super-negative, and rubbished the concept, adding that it's easier to just do it on a machine.

I was a bit shocked, since he use to be a maths teacher in Lebanon, that he could not appreciate the apparent beauty of Mental Calculation and it's benefit's.

As he continued to bag it or curse it, I had to interject my comments:

"You're so negative

 Wake up"

My voice began to get louder, it was like I was releasing 30 years of pent-up emotions of my father not accepting my right-brain visual mathematics, and I began to repeat myself, and commenting further:

"Wake Up!

You ought to be ashamed of yourself. You call yourself 'Istez' (teacher) yet you rubbish this new maths.

The Universe is trying to show you that this is the New Learning and that children like it. Channel 9 is showing you the happy and impressed parents of the students learning this.

Wake Up," I said louder again.

Then he freaked and got upset and swore in Arabic, and paced the ground, Mum told me to stop. (in our Lebanese culture, no matter what, whether the father is right or wrong, you do not challenge him).

Dad turned off the TV, turned off the lights, and we suddenly were sitting in the dark.

It seemed to be the only way he could deal with his denial, was to shut everything off.

But mum was sitting near the switch, and turned it on, somewhat irritated.

Dad and I started our verbal combat. He kept on calling me a failure etc I responded that perhaps he is jealous –

Mum insisted that Dad go for a walk outside, which he did with a black cloud around him, cursing me. It was like he was going to have a heart attack _ _ But he only lasted one minute, and came back into the arena for more. This time he threatened me to pack up and get out of his house or he'll shoot me, knowing full well that he knew that I knew that he did keep a rifle in his room.

Mum tried to calm him down, and waved a hand to indicate that I better wisely get downstairs and end it, as he was furious and unpredictable.

I did not respond as I could see madness and fear in my dad's eyes. I saw him as a child too like he was a victim of his father telling him such horrible things of failure etc and that Dad was projecting his wound subconsciously onto me.

So that glimpse or insight calmed me down and so I went to my room to pass the time, reading Lucas Sequences and Brown Landone on Teleois 1, 4, 7 pyramid codes. My father, 20 minutes later, crept downstairs, ignoring me. I heard him locking the office door as if he was afraid of me entering in there, I saw his stern face as he snuck by, but he did not dare to look at me as I was sitting in bed reading. His presence downstairs was a manly or tigerly tact to assert his dominance. I knew not to talk or react, just to observe, and stay quiet by keeping to myself.

So that was the end of that Saad Saga...

I went to sleep for a short while, as I was tired from 2 weeks of Sydney lectures and this debate, but tossed and turned a lot thru the night.

Not able to sleep, I got up at 1.45 am to have some water and decided to write my thoughts out on these pages.

I remember, this morning Sephira rang me on my mobile, and she thanked me for cancelling the Melbourne leg of my tour and for deciding to drive back home to see her at Murwillumbah, as we want to marry and live together.

She thanked me, and squealed in love in that special voice of hers. She said quite wisely:

"When a man is on his death-bed he is unlikely to say that I wish I had more time for work or for more money. No, he will say I wish I had more time for my children".

That was so beautiful, and at one time this afternoon, pre-fight, I said to Dad, after he was again rubbishing my work, about Sephira's wisdom. It must have affected him on a deep level.

My Dad, unable to understand Sephira's wisdom story, asked me to listen to his story.

Dad often told us Parables, it was one of the special things that he did do to us 3 children (Robert my brother, 1 year older and Faye my sister, 1 year younger, I was sandwiched by age in the middle), was to recite **Lebanese Parables**, making us regard him as a teacher...

or so you would think...

So he came back to my story, with the following negative story, it's punch-line was another reference to my failed life...

The story goes like this:

An older man had 3 sons and brought them 3 to his home. He asked them each to travel to a place where they had not been before, at their father's expense, for a period of 6 months, and then come back and report to him of their worldly choice and experiences. The 1st son came back and said he started a business and that it began to prosper...

The 2nd son also started a project and it too prospered quite well.

The 3rd son went into Nature and said nothing had happened of any significance, he said to his father, but he did see an unforgettable event where an eagle had stalked on a lost lamb and ate it in part, followed by the other eagles who subsequently shared on the feed. When the birds had had their full, they left the carcass. The son then noticed that foxes came and ate the rest of the flesh, getting a free and easy feed.

The father then retorted to his 3rd son, "My son, I want you to be the eagle, be the successful hunter, not like the fox who feeds or scavenges from other's victories"...

I actually liked this story, thinking it very clear and wise, but realized very quickly that it was another smart and subtle attempt by my father to analogize me to the scavenging fox.

Imagine stories like this all your life. Imagine having a multi-millionaire father who lives for money and can not spend time with his own Son who may be a mathematical genius.

My father asked me this 'avo, pre-fight, "how well did I go in Sydney, how much money did I make."

I said about $8,000, mainly from being 4 days at the MBS Mind Body Spirit Festival, as I was a guest speaker on main stage 2 of the 4 days there and sold my dvds and books on mathematics like hot-cakes...

But again, my father rubbished that success, saying it's all a flash-in-the-pan, meaning that the success will not last nor repeat.. so from comments like this, your judge, your highness when my father died from his sudden heart attack after we had this math solution debate, it was from his own mind of failure and defeat, he had failed to love his own son, me, it was not my problem, it was his own mental sickness, your worship, I just had to remain strong and tell him most strongly to "Wake Up" to "stop this Negativity" for God's sake, it was his deep anger, that killed him. It was so inflamed, by him, not me. I have to thank him for making me a stronger person, as I now refuse to tolerate any negative comments to enter my inner landscape. I value who am I, I love my **Mathematics for the New Millennium** and will continue to do so to share my bliss and wisdom to all children of the world, in all languages of the world. It was sad that I had to learn to walk away from my own father to merely dodge those darts he threw at me. My only error on that day of his death, is that I lovingly wanted to see my mother for a belated mother's day, b4 I drove home to see my Sephira... in my father's parable, I am The Eagle, I have the strength, the inner strength to repel back to the source, any negativity directed to me. It's called the **Law of Boomerang** or **Return To Sender**. My father got what was due to him, and he was born a sour seed that grew into such a tortured soul. It has been painful for me to watch a man live in such negativity. I am a father also, I encourage my teenage daughters **Mingkah Jain Sun** and **Aysha Jain Sun** to follow their bliss or passion, I support them, yet my father hated my creativity, and tried always to stifle me with his narrow wit. He is better off dead, your honour, his time was due, I was only a catalyst...This reminds me, I abhorred and was truly in shock, the way my father, in my teenage days, sacked **Aboriginal people** who turned up onto our building sites enquiring work from our newspaper ads, and commenting to me that the whole black race should be exterminated! I never forgot his racist remarks. I see now that his wound was inherited from his own father's wound, for many reasons... that a poor life as a peasant with the eternal hoe in his hand ekking out a miserable existence from sunrise to sunset prevented him from becoming a doctor or a learned man, so he trekked to other side of the world where rumours abounded of wealth and good-living, so he came to Sydney with a Dream, at the expense of his father cursing him, and made his millions. As we 3 children grew up, he said all this wealth was for his children, but somewhere along the line, he forgot the true meaning of his pilgrimage to a more civilized land.

He can not see that I am his **mirror**...

He was a Teacher of Knowledge and Mathematics in the Mountains of Lebanon. I am a Teacher of Mathematics in the Mountains. He left his Country for a better life. I left my Family to travel the world. He always wanted to be a Doctor. I am a Complementary Physician. He changed his family name from Yazbeck to Saad and I changed my name from Mr Collin Saad to plain Jain, or Mr Sad to happy Jain. It's obvious I am my father, whether I like it or not!

Thanks, Jury, that is all I have to say.

ps 1: As I switched the light off, I was looking at the time on my mobile phone and just for a flash of a second I saw it said

"3.14"am

(nb pi = 3.14 which is a circle/square relationship).

15-5-07, the same morning 8am

ps 2: It's also interesting that whatever we judge or condemn or react to, we feed it, and therefore create it in our life. Here are some thoughts regarding Prof Hanley of Victoria. I have blacklisted him to the international Vedic Maths community on the grounds that he had written some books about mental calculation and not once acknowledged the Vedas or the Master Bharati Krsna Thirthaji. All he had to do was state that it's possible origin was from India's past, and who knows, it may have preceded this and been

Atlantean knowledge or even prior to this. But as a scholar he is required to state his sources of research, unless it was pure self-discovery! Whatever the case, I warned my parents as they were about to watch this "Math Solution" on Current Affair, that he was an imposter, a thief, for the above reasons, and realized that I was very disturbed by his presence. Perhaps seeing his success on TV and releasing this knowledge out to the world made me somewhat jealous, and therefore contributed to my raised voice against my father, I don't think so, but I don't rule out the possibility of this being a factor, as we are all students of the Mind, all here to learn the truth of who we are. So even though Prof Bill Hanley, the man I judged for 10 years, is an imposter, he still has his divine place / role to play. The fact that he appeared on TV for 5 minutes on prime time on Channel 9 around 7pm was wonderful, that mainstream people had the opportunity to see a brilliant style of modern mathematics (that is really from the ancient past). So are we not to praise him for his efforts. We must cease this cycle of blame and condemnation. Here I am condemning my father, because he condemns me, and I condemn Bill Hanley. How primitive and childish are these emotions of negativity ruling our Minds and therefore our Hearts. This is quite bizarre, yet beautiful, this raising of our consciousness, by learning to hop off the wheel of blame and condemnation, to stop judging.

Must I now support my enemy and join forces with Bill Hanley to package this ancient Knowledge, (not Indian Knowledge) and create the universal mathematical curriculum for the advancement of the planet, a system of number study that embraces all cultures, now known and coined by me as "Global Mathemagics".

And did I create this Episode. How timely was it, that I walked from downstairs to upstairs to have a kindly cup of T with my mum and dad, and there on TV is the very enactment of my life's work, just b4 I travel a 1,000 kilometres home and I had deep down wanted my parents to know what it is that I do or have done for the past 30 years. And it just appears... Is that **co-incidence**? I don't think so. Are we gods in the making, learning to see god in all, even in our enemies. The error of the human condition is that we create division between good and evil, yet all is one, all is divine. How brilliant, that in a crack of Time, my **quantum Infinite Mind** was able to project to my loved ones the very reality that I desired for them to see, at the right moment in the right time. It was like time slowed down for the gears to get into order to access the vibrational akashic spirit records, bit like in the **Matrix movie**, when Keanu Reeves is being shot at, and he is dodging the flurry of bullets heading towards him, all in slow slow motion, and effortlessly dodges each one... so in my case, time slowed down and this event happened, a curtain was drawn open for a show to begin.

Why love and understanding was not created, I do not know, that will be the subject of my next enquiry and thus next article.

I guess deep down, since I too am now a Father, I would like my own biological Father to at least say: "Son, I do not understand the mathematics that your are passionate about, but bless you my Son on your Journey, may you succeed and have a happy life. That is all I would love to have heard from Dad b4 he died in his own mental metal poisons" your Honour, your Judge.

When you think about it, what is a **Heart Attack**! It is a **Love Attack**! When we are unable to **share** Love, unable to say kind and sweet words, we are really attacking ourselves, seeing it projected out in others who **mirror** it back to others for us to see. I therefore bless and thank my Earthly Father Peter Boutros Saad Sakoua Yazbeck for being a teacher to assist me in **boomeranging** back other entities' negative projections, that has made me stronger and sharper in wit, has made my heart full of love to only accept and invite true love into my life, that anything less than this frequency of love I do not absorb. And that our error in thinking is that we have created the concept of Good + Bad, a lover and an Enemy.

It's all in the Mind, a **FrankenStein** as warped as seeds genetically modified. There is no right nor wrong, no enemy, and therefore no angel. As soon as you have aligned your neurons in this duality, you will surely create that movie. My father taught me this, I am sovereign unto myself, that I believe in myself, in my Vision, no matter how much he may have tried to ridicule my Journey, to humiliate and de-value my life, I have learnt to be Discerning and Empowered.

"And I did not cry at my father's funeral", your Honour, your Judge, "for he was just a shadow of my Mind, a cobweb in my Consciousness that needed flicking away like some darkened flackened string of annoying snot, a bit sticky that it took many flicks, a bit strange that he claimed to be a mathematician in the mountains of Lebanon and you would think he would have embraced my number revelations and superb original digital compressions, a bit sickly that he could speak 4 languages, like Aramaic, the tongue that Jesus spoke, so you would imagine that he was a learned man, a pundit, a bit disturbing that he was a business tycoon unable to share his wealth to his own blood and bones, a bit tricky in that he was my father, but so what, so what, everyone IS my family! That is all I have to say," your Worship.

My girlfriend Sephira is 7 and a half months pregnant to another god, and I will marry her and adopt and love her Star Child. Sephira is the wise one here. Remember her words: "When someone is on their Deathbed, they don't wish that they worked more, they simply wish that they had spent more time with their family". It's so true, you will invariably be crying for more love, not more money, for more time spent with those who you care for, and who care for you. This is the Cosmic Truth. Sephira is the Goddess of Love, and her Love is so magnetizing that I am drawn to drink and totally absorb her Mind and Heart and **Yoni** Essences and Bloods. Drop all Judgment, love your Enemy by realizing there is no Enemy, it was all Illusion, and begin to

Live Life

Love Life

Live Love

This is the true **relation-shift**.

Sephira's love has slayed all my ambitions, as I realized in her beautiful moans and squeals of ecstasy that her Love is the Goal, her Existence that exudes love into the pores of my skin is so total, so uplifting, that all my teachings of Fractality, Compression, Number-Harmonics are meaningless if I am not first aligned in Love. As we merge, we will teach together, the magic of the Now, not The New Age, rather The **Now Age**.

Walk away from anyone who begins to talk about the future date: 12-12-2012, it's all **Mayan** Madness based on Fear and therefore a lack of Love, with imaginary or invented Days Out Of Time. If for some reason you do not see me on the fake lecture circuit where **egos** collide and international contractors are foxes in disguise, it is because I am resting on Sephira's lap whilst she is tickling my spine and if the phone rings and I do not answer it is because we are adoring one another, exploring one another in the deepest possible configuration of the most holy of holies **69**. When love is in my

Heart and in every cell, only then, when the Divine Man is **Imbued** with the Divine Feminine can I be whole, can I take my first true step on Earth, to claim then teach this galactic mathematics of the moon... but is the moon real or is it a hollow metal sphere brought in from our solar system when Vulcan or Maldek blew up and became the Asteroid Belt between Mars and Jupiter, which disturbed Earth's original **360** day cycle and planetary path around the sun and to correct the wobble to a current **365** days, some shape-shifting, gold-eating, master **gene-splicing** aliens whose **dna** is reptilian, dragged or towed this large rock into our orbit, using a great **Imagineering** Mind and Mathematics to solve a problem, so the legend says, and if this is true, what would the women of the world think all this time worshipping the moon as a goddess of love, hmmm, what a major dilemma. You see, it's all a story, history, herstory, it's just another of a thousand creation myths, and in fact they are all true, because whatever you think, so be it...

So right now I am on my merry way to see Sephira, coz I think she is The One. The pain of being separated from her these last 2 weeks is unbearable... I have crazily cancelled my tour to Melbourne, to get Sephiralized, to return to my true Sun Centre, not to some fragmented dust-particled false moon-glowing imposter planet... to live a life based on the fact of Love, not a fairy tale.

Thus, in this account, this fable

Jain 108 Vs

P.Saad Vs

B. Hanley Vs

Sephira

there is no question

Sephira's Love is the Superior force:

The ultimate Divine Proportion.

I realized this to be confirmed when I learnt of her natal surname:

"Sephira Phi-Lips".

Everything she speaks or that which is formed from her lips, are the sweet Phi or Divine Proportioned golden words, golden mouthed poems. Sephira has taught me one special thing, to be golden mouthed which means only to speak little, and when you do need to speak, let it be the Truth. Only the truth.

"This is the reason why my father died of a **heart attack**, your honour, your worship, he was really seeking Love, and he Feared me, he Feared Life. He was a Lie, an illusion, a prisoner of his own limitations, a shadow, he was not a darwinian survivor of the fittest, he was essentially weak, he spoke ill of my new love Sephira, he spoke ill all of my life, he condemned the aboriginal race, he condemned my mathematics because he was already condemned at the core, he was devalued by his own father, so he never had a chance to survive in the atmosphere of love, that is why he got lost in his dollars, he was already half dead, I only stirred his anger that night, he was already wounded, I merely pushed him over the cliff of his own nightmares, there was no thinking nor plotting, it just occurred as a natural event, a calling, and that was the end of that Saad saga, the phantom had gone, the master/slave episode faded. He could not look me in the eye and simply dropped dead. How beautiful! He could take no more. I shocked him, and he slipped thru the crack.

I have no regrets, it was merely a service to progress life and accelerate the karma, and arrive to a new destination beyond that shedded skin.

Jain 108, Kirrawee.

Art of Jain, 1979, "first Self Portrait of Collin Saad"
(pronounced as "Calling Sard")

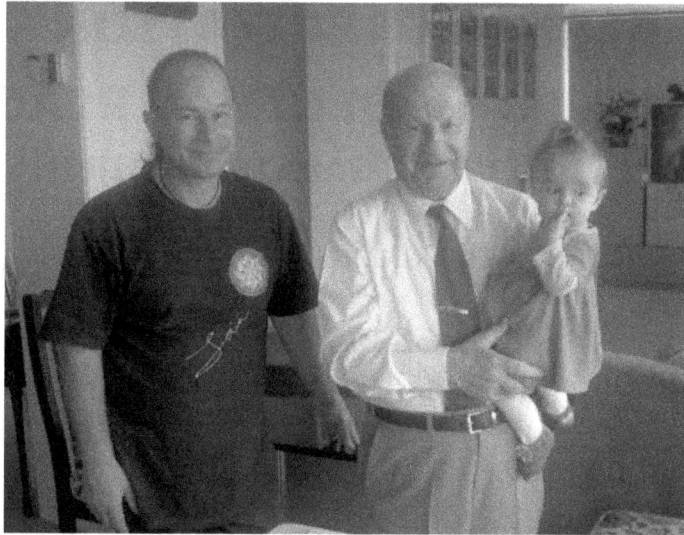

**May, Sydney, 2008,
from left to right: Jain, Peter Saad, Galatea**

**Energy Rainbow Burst of my Mum: Yvonne Khoury.
Compare this to the original one above
NB: these photos have not been photoshopped. It is pure energy.**

SEPHIRA RECITATION
OF PHI TO 50 DECIMAL PLACES
(ON THE EVE OF GALATEA'S BIRTH)

(this is what is written on the diagram on the following page:

"1.61803398874989484820458683436563811772030917980576...
for GALATEA 108 26th July 2007 (ne 27-07-2007)

These 50 decimal points of the Phi Ratio 1:1.618 were memorized fully during Sephira's birth labour on the eve of birth, with Jain 108 reciting it as well. We both had turns to recite these 50 decimal digits from Memory, not having known this sequence or order except up to 9 dp! Especially for Sephira, it seemed like a door or portal in2 Hyper-Space had opened up, and she entered a dimension of All-Knowingness, of Clarity and Oneness. We were just sitting on the floor lounge in our living room (near the in-door birth pool) at 777 Left Bank rd, Mullumbimby Creek, far north NSW, 2482, Austalia, were playing this recitation game playfully as in contest to outwit one another in Memory Power, and were both surprised that within 10 to 20 minutes we had fully learnt it, like idiot savante geniuses. We had definitely tapped in2 some altered state of Higher Intuitive Mind... and soon to follow, that twilight night, the birth Journey of Galatea Prana had begun".

...as remembered and dutifully archived by Jain 108 and written on 21-12-2007, just for the record.

1·618033988749894848204586834365638117720309179805 76...

Upon the Eve of Birth of GALATEA
(né 27-07-07) @ 777 Left Bank Rd...
These 50 decimal places of the Phi Ratio were
memorized fully during Sephira's birth
Labour, on the Eve of Birth, with Jain 108
reciting it as well. We both had turns to
recite 50 decimal digits from Memory, not
having known this Sequence or order, except
up to 9 dp! Especially for Sephira, it seemed
like a door or Portal in2 Hyper-Space had
opened up, and she entered a Dimension of
All-Knowingness, of Clarity + Oneness. We
were just sitting on the lounge-room floor (near
the indoor birth pool), playing this recitation
game playfully, as in Contest to outwit one
another in Memory Power, and were both
surprized that within 20 minutes we had fully
learnt it, like idiot-savante geniuses. We had
tapped in2 some altered state of Higher Intuitive Mind ... & soon to follow on that twilight
nite... the birth of Galatea... written by Jain.

Sephira Phi-Lips Recitation of Phi to 50 decimal places!
Whilst in the birth pool, hours before the birth.

Sephira with Body Paint,

Body Art by Jain 2007

(Digital Photo by Jain, version 1)

version 2

(just another perspective of the pregnant goddess.

Nb: this digital photo has not been photoshopped, it actually appeared like this, as explained to me by Prof Harry Oldfield of the Uk, the creator of the first auric photography. He explained that it was most likely created and affected by my own energy field!).

cHAPTER 7

"JAIN One O Eight EARTH-HEART
In The COURT Of Mind"
By Jain 108

(May 2007, Sydney)

Reason for Court Case:

Jain 108 EarthHeart and **Miss Phi-Lips** arrested and charged for "Making Love and Not War" @ the **Talisman Sabre War Games** @ **Shoalwater Bay**, near Rockhampton, mid coast Queensland,

June 2007.

[Talisman Sabre 2007 Involves 26,000 US and Australian troops over the period June 10-30. The exercise is taking place mainly at Shoalwater Bay in central Queensland and in the Northern Territory. The exercise also involves the use of civilian airports, including in Sydney and Brisbane, and Amberley air force base near Ipswich. Talisman Sabre also involves US nuclear-powered naval vessels.

In 2005 and 2007 a number of protesters were arrested for carrying out non-violent protests within the entrance to the Shoalwater Bay Military Training Area. Some groups have organized a 'Peace Convergence' in protest of Exercise Talisman Sabre's environmental and social impacts and calling for an end to the war in Iraq, and Australian involvement with US-led wars. The weekends of protest around Shoalwater Bay involved a variety of actions from peace walks to film showings, to vigils and other non-violent direct actions. More than 1500 people marched in opposition to the war games at nearby Yeppoon on June 24, 2007].

Judge: Can "**Collin Saad aka Jain 108**" please come to the stand.

"**Jain 108**" comes to the stand.

Judge: Can you please state your name for the Court's records.

Jain: It is "Jain One O Eight". It is not "Collin Saad". Deed Poll law 23 years ago made me sign and declare that I could no longer use my birth name of "Collin Saad", that this name was invalid, so please address me as "Jain 108 aka Collin Saad", as in "also known as Collin Saad" or once known as "Collin Saad". But I am not "Collin Saad aka Jain 108" but rather the reverse, I am "Jain 108 aka Collin Saad".

Judge: Mr One Hundred and Eight, is that correct.

Jain: No, it is Jain One O Eight, not mister, but Jain.

Judge: Ok, lets try again. "Jain as J-A-I-N" is your first or Christian name, pronounced like the girl's name "Jane", is that right?

Jain: Yes.

Judge: And your surname is "One Zero Eight".

Jain: Yes technically, but casually, No, it's "One O Eight".

Judge: That's what I said, "One Zero Eight".

Jain: No, It's "One O Eight", You said "One Zero Eight". It's not said as Zero, it's said as an "O", as in the letter "O", to replace the number Zero.

Judge: Yes, but if your name is a number "One Hundred and Eight" you can not use letters of the Alphabet as in "o" to replace the number "Zero".

Jain: Yes I can, it's my name. I don't choose to say "One Hundred And Eight", it's too long, plus, my name is meant to be a number as in "108" only, not spelt out as it is now. My name is "Jain 108" but your government computers won't accept the number "108" so I have been forced to write it or spell it out which has complicated the issue here. I am "Jain 108", just a number.

Judge: Ok, so you are not really Jain One O Eight, you are really Jain 108.

Jain: My surname is pronounced "One O Eight". Anyway, Zero is not a number, it is the infinite void.

Judge: Ok, "Mr Jain 1 – 0 – 8", it is. Have you registered this name by Deed Poll?

Jain: No, I have no need to. It's just a name, or should I say, a Number! I originally had a single Deed Poll registered name, as "Jain" in 1984, I believe also that I was the first person in Australia to have a legal deed-polled single name, which meant by law that at my birth "Collin Nicholas Saad" had to be sacrificed and was illegal to use. But I must confess, having a single name has been very difficult in the last 3 years, travelling around the world, having only one name "Jain" and harassed by all the airport institutions who require 2 names on their computer databases, a first name and a second name, so after 23 years, I have chosen to go standard and have a normal name with a surname, just to make things easier. So when all the Institutions demand that I need a second name or a surname I just had to think, on the spot, of a better name than what they were calling me or had me listed as.

Judge: What were these names that you are now rejecting?

Jain: Like "Jain Jain", a doubling of my name, which is actually quite common in my Arabic tradition for a man to be called Mansour Mansour or Farouk Farouk or Saad Saad etc it didn't matter so long as something was entered in that second field, even "Jain X" would suffice. And to complicate matters more, my passport is written only with a single name Jain.

Judge: It shows on some older records here that "Jain" was your surname not Christian name, please explain.

Jain: Oh yes, my first full name was Mr "Just Jain". I would be explaining that I only had just one name, you know what I mean, I would say that I am just "Jain" so they thought that my first name was "Just", and even though I protested, what the heck, I just surrendered to that title Mr Just Jain. Though sometimes, I would alter it and say "Plain Jain."

Judge: And what is this signature here, I have three "Js".

Jain: Oh yes, during that same time in the late 1980s, other institutions would demand a middle name, so on the spot, I would say "Jain Jain Jain", my middle name being "Jain" again, so that epithet stuck as well, and became bit of a nick name. My close friends started calling me "Triple J" which was the name of a alternative and still contemporary radio Sydney station.

Judge: Though now here in this Court, it seems to be quite a complex matter to list all these past titles and now call you a number 108.

Jain: No, not at all, your Highness, it's quite simple, really. My surname is just 3 short syllables: "1 – 0 – 8".

Judge: Lets move on. Can you tell the Court your original birth name, just for the record.

Jain: "Collin Nicholas Yazbeck".

Judge: Oh, I have here "Collin Nicholas Saad".

Jain: Yes, that was my name for 27 years but I learnt at the age of 27, when I went to a Sydney office at the Department of Births, Deaths and Marriages, to do the legal Deed Poll in 1984, from "Collin Nicholas Saad" to "Jain" that they could not locate that name. Which means that my official school name "Collin Saad" did not exist! So technically, I had never existed, if you go by the books. Somehow, after much enquiry, phone calls and more visit's, I learnt that my father "**Peter Saad Yazbeck**" put down my original birth surname of "Yazbeck" as "Saad". You see, in Lebanon, he was born a "Yazbeck". In those days, they did not even record his birthdate, he still does not know what day he was born. When he decided to migrate to Sydney, from Lebanon, to have his family, he wanted his future children to have an ozzie name, so he changed his name from "Boutros (Peter) Yazbeck" to "Peter Saad". Now, "Saad", (pronounced as Sard, which means "Good fortune", or "Lucky" was my father's father's first name, that is, my grandfather's name being Saad Sakoua Yazbeck, whose middle name "Sakoua" means "Sleepy Head" and became his nickname and also got tagged onto my father's name, and also my name. That is why as a child, I was known as "Collin Noula Saad Sakoua Yazbeck". The "Noula" bit is the Arabic name for my middle name of "Nicholas". So dad thought that "Saad" was a cool ozzie name, little did he know that ozzie people could not pronounce "Saad" as "Saar-id" and people at school, like my school teachers during roll-call time, called me as "Collin Sad". So imagine all your school life being called "Mr Sad". Now you know why I had a legitimate and genuine reason to change my 5 complex and unpronounceable Arabic names to a single one simple name Jain.

Judge: But your first name of "Collin" you said it was unpronounceable!

Jain: Well, yes, in the sense that my mother never called me "Collin" as you would pronounce it, she always called me "Calling". You know when you hear someone's answering machine, and they leave a brief message, and it often ends in, "Thanks for calling". So I was called "Calling" which is ok, I have no problem with that, but I am either "Collin" or "Calling", it should be black or white. Anyway, I did not like this "Sad" bit, I just wanted to be happy. I think I was the first person in Australia to have a single name, not that this is important, and nor did I persue this for this ulterior motive, I had a genuine cause or reason.

Judge: It appears that you did pioneer this name change, and also to have the first surname as a Number!

Jain: Though, we all know of the rock'n'roll star "**Prince**" he does not even have a written name, he is actually a symbol!

Judge: I believe so. Anyway, we are wasting time on this issue, we need to move on.

Though, one more last question. Is there any meaning to this number 108?

Jain: Yes deep meaning. It is the secret and hidden vibration or pulse of the numerically compressed Fibonacci Sequence. We have been told for thousands of years that there is no distinct pattern in the Living Mathematics of Nature expressed as 0-1-1-2-3-5-8-13-21-34-55-89-144-233-377 aka the Fibonacci Sequence, not that it belonged to this Italian man Fibonacci, it was known by the Egyptians, Vedic cultures etc as well as in the pre-ante-diluvian times aka **Atlantis** and **Lemuria** civilizations. I have devoted 30 years of my life to full time research on this number code of nature, and I believe that I have cracked the code, bit like the Da Vinci Code material or bit like Russell Crowe/John Nash in "**A Beautiful Mind**" and this is my gift to the world, not that I am schizophrenic like John Nash, the main character in this movie, but lets say that I have x-ray eyes and can perceive the invisible inherent order amidst the overwhelming visible chaos. I have exposed a secret mystery, an underlying current or pulse in the universe that the ancient Vedic culture hid sonically in the famous prayer for enlightenment known as the **Gayatri Mantra**. It's all connected to the 108 rosary beads that they all clutch, the 108 Upanishads, the 108 bricks in their altars, the 108 number of stitches on an American base ball, or cricket ball, I forgotten which ball, but count them and see. I count everything. 108 is an Anointed Number that taps into the bridging of the Atomic world to the Galactic Memory, call it Microcosm/Macrocosm but all you need to know is that these Fibonacci numbers when translated into art create the nautilus shell spiral in 2-dimensions, which when revolved becomes the ram's horn in 3-dimensions known as the Phi Spiral vorticities, which when revolved again into the next dimension becomes the shape of all blood platelets known technically as the tube torus domain. The Mathematics of the Soul is the key to compression and we call it a proportion known as **1:1.618033988...** resonant in our **DNA**. That is why every living protein in our bodies is based on the pentacle as the pent molecules is phi ratioed. It got demonized so you fear this Mathematics of the Soul, but all you need to grok is that this maths is in flowers, it's in the distances of the planets from the sun. We call it "**21:34**" this is the signature of the universe: 21 is to 34, the counter-rotating whorls or spirals of the sunflower floret. It's nature. So that is why when my students say something complimentary regarding that things are flowing and well, they say it's "Twenty-one Thirty-Four" (21:34). In all of this is the 108 mysteries, and we call it holy or blessed by calling it "**Shri 108**".

Judge: So you elected this Number Name because you consider yourself Holy or Blessed?

Jain: No, your worship. It just happened naturally. In Nov 1984 I was flown over to Kuala Lumpur in Malaysia to meet with the Prime Minister's chief official, at their bequest, to assist them with understanding Vedic Mathematics or **Rapid Mental Calculation**. 10% of the population there is Indian, but overall, in the daily newspapers, on the front page, it was saying that Malaysian students get a big "F" in mathematics, which means that overall, this culture just did not grok the British curriculum, and so they consulted me as an expert, how to translate numbers into art to join the left and right brain cortexes to make maths more visual and easier and fun to grok. I then flew over to India and showed the top mathematicians there, working at Info-Sys in Hyderabad, the magic of 108 disguised in the compression of the Fibonacci Numbers, and they were bowled over. As the year rolled by, back home, I was receiving letters from these officials and mathematicians from 2 countries, and they were naturally addressing me as "Mr Jain 108". I still have these original letters to prove this, should the Court need this as evidence. And on some letters, I was known as "Sri Jain 108". Thus it remained as a obvious and natural choice. That lecture I gave on 108 must have been pretty good, as they are still talking about it. The school teachers and students whom I also taught in Kolkata and other cities were quite surprised that I, a westerner, was introducing to them their own lost heritage of rapid mental calculation and Number Theory based on 108. You know what I mean. Here is a culture with a billion people that mindlessly and devotionally worships the holy Number 108, but if you ask them why, they do not know why. They just blindly do their daily 108 chants of the Gayatri Mantra without any thought of the mathematical derivations or origins of this number. So along comes "boy Jain of Oz" ('cos some people think that I am a girl, so people do call me "boy Jain" to distinguish me from "girl Jane") and blows them out about the intricacies and intimacies of 108. I did not plan this name, it just evolved or revolved. In my life, I have been a pioneer of new ideas that get adopted. I am thus calling this new syllabus of material as "**The Mathematics Of The Soul**".

Judge: For the record, can you list some of the works that you have pioneered.

Jain: Yes, my best achievement, is that I, an Arab or westerner, gave to the world the first dvd on Rapid Mental Calculation, which means, I beat the Indians to putting out this rare and lost material on Vedic Mathematics in it's visual form. It's probably what has made me famous, and the reason why the name 108 has stuck.

Judge: Good, and another example, if you please.

Jain: I have several other innovative or original examples. In 1984, as discussed in this Court, I was the first person to have a single name.

Besides this, and around the same time, I came up with a brilliant and ingenuous One Word Poem. I believe it is the most profound, simplest and cleverest poem in the 21st century. I have seen it all over the world during my global travels, and many people adopting it as logos and used on stickers and decals and titles everywhere. It's called "**EarthHeart**".

Judge: I don't get it.

Jain: It's so simple, if you take the One Word: "Earth", and write it many times, in a circle, as in a necklacing of the same word like beads joining, "Earth-Earth-Earth" and read it again and again with your eyes squinted, so that there are no spaces in between the words, "EarthEarthEarth", it now reads as "Heart Heart Heart etc" or by removing the spaces: "HeartHeartHeart". It's quite brilliant and simple. Yet it is very profound, implying this deep connection between Earth and Heart. I resonate to this symbolism since in 1984, I was stabbed in the heart chakra and managed to survive, and subsequently developed a "One Breath into the Heart" technique, now known as the EarthHeart Meditation, and is an integral part of learning this Galactic Mathemagics that I teach, that we learn from the Heart, breathe from the Heart, so as not to remain in our Heads. You can read about this in two of my green books on Phi. EarthHeart is about finding our true centre, and from this authentic space that connects us all, we can learn the truth, we can then understand higher principles that the true value of pi (which I call **JainPi**) does not equal the disharmonic book version of 3.141592…, but is based on the radical or square root of the golden mean being precisely 4 divided by the square root of Phi giving the **true value of Pi as 3.1446...**

EarthHeart is therefore the **Temple of Mathematics** that is giving the next generation the chance to access the Spirit Records, if it be the will of the People and Politics, which is as you know it is more than a tug-of-war, it's a Star Wars game.

Isn't that Beautiful. I am so proud of this one poem. If I was to die tomorrow and be remembered for one thing, this is it. What an amazing connection between the Earth and the Heart. Actually, there is more to it than meets the eye. If you squint your eyes again, there is another infinite recursion in this Earth-Heart literature. It actually reads: "Hear The Art, Hear The Art, Hear The Art" or removing the spaces: "HearTheArtHearTheArtHearThe Art".

You will see such puns as this on buses and websites, it's everywhere, and it is not vain of me to state that I am the seed of this. A pioneer.
(see Appendix 3, on page 181, for the full Poem on EarthHeart).

And just to conclude, I have created many other clever titles, one that is relevant to this Court, is the ecosophical slogan now seen everywhere is: "**Inner Peace = World Peace**". I painted this on my JoyVan in 1987 and yes, I am proud to say it has gone worldwide.

Also, around this time, I was pioneering another artistic movement aka Fluorescent Mandala Banners, from the time of 1984 to 1987, in Upper Dingo Creek, after the time that I got stabbed thru the heart chakra, I was rehabilitating in nature there, living as a hermit for 3 years. I introduced the concept of painting sacred geometrical thangkas or mandala banners based on magic squares and 108 harmonics and painted only with fluorescent t-shirt fabric paints. I was alone doing this, it had not been heard of, and 10 years later, it was all the craze in Byron Bay, far northern New South Wales. They looked great, as they glowed under those blacklights and was a two-way artform, as they appeared to be different geometries under the blacklights. They were very popular at the outdoor trance-dance or doof parties that were raging at that time. But my paintings were not to support that droogy or drugged-out entheogenic culture, my work was at a more higher or sensitive frequency, in the sense, as a Keeper of the Records, I was painting it and putting it out there.

Judge: Ok, that will do for now.

The Court will go for a break now and resume with a testimony from Miss Sephira Philips.

Judge: Ok, I call Miss Sephira Philips to the stand.

Phi-Lips: It's pronounced "Phi-Lips" not "Filips". It's spelt as "Phi" and "lips".

Judge: But if I was to read "Phi" I would read it as "Fee" and therefore as "Miss Fee Lips". Is that correct?

Phi-Lips: No, my name is Sephira Phi-Lips, your honour, and I am very passionate about this, "Phi" as in "Fe – **Fi** – Fo – Fum" (she accentuates the Fi). Phi is the Divine Proportion as seen in Living Nature. Phi Lips conjures the ability to speak through a mouth that is in the Divine Proportion which is just the pure mathematics of Beauty. It's about being **Imbued** by the 108 frequency aesthetics of proportion and ancient mysteries. It's suggested in my surname, "Lips of Phi-Lips" the art of speaking, Golden Mouthed.

Judge: How long have you had this name?

Phi-Lips: Only 6 months recent, actually since I met Jain, and I was so profoundly captivated by his Knowledge on Phi, that I am still imbued by the concept of being **Engrailed**, to grok that the Holy Grail that we all seek is really the Ace of Cups, the abundant cup that overflows, is the 108 that Jain, my Cosmic Beloved, is on about. I love that word "Imbued", it just sizzles me! I love Jain. When I drink Jain I drink in the Phi. I just want my name to embody that, to become that, to express that power of words, thus, I AM "Miss Sephira PHI-lips". I want to have his babies and nurture him with whole foods, and unconditional love, massage his soul with eternal joy and happiness because we are co-living, eco-living withIN the frequency of 108 that engulfs us, with our worm-farms and our daily rejuvalac which constantly is rejuvenating our evolving love on this plan plane planet.

Judge: We need to know for the Court's records, are you in a defacto relationship with "Mr Jain 108"?

Phi-Lips: Yes and No, No we are not in an earthly relationship but yes we are in a spiritual and cosmic relation<u>Shift</u> connected thru our Adoration of Phi. This makes us Engrailed. Jain is my Beloved. He is teaching me all about the 108 Mysteries. It's so profoundly abyssmal and fractal.

Judge: Abysmal as in deep, but what is this fractal?

Phi-Lips: Fractal is not a fraction, but rather the opposite, about wholeness and infinitude, also known as "**in-phi-knit**". So during those public war-games, when I took Jain into my Body, my womb, what you call having Sex, it was a sacrament to Mother Earth, an honouring of the Creation of all Life, that we are all One and intrinsically linked to one another, that is why we could not celebrate your wars and destructions of this planet, it was important in that moment to send Love Vibrations to all of our brothers and sisters. And because this area is a nexus or juncture point of the world grid, it means our Love would radiate to all areas of the world.

We have committed no Crime, we teach only Love, for that is what we are.

Judge: Yes, but by being in that vicinity of war games was endangering your life and others.

Phi-Lips: I have nothing to lose. I have no family lineage, I am the end of a lineage, now that my mother has died 5 years ago, and my father died when I was 9, so at 33 years of age, I live on the brink. That is why my child with Jain has a new name, a name that is not mine, and I cant take on Jain's name 'cos as you have established it's either a single name or a number, and in either case, my daughter who is now 4 months young, has the surname of "Galatea Prana", which is the Life Force. And as you gather from all the events that have transpired, that the second child that I now bear in my womb, was conceived on this controversial act of making love and not war at the Rockhampton Sabre War Games. So I just want to make this point very clear, and I challenge you all, that thru our actions of making Love we have created sacred Life as opposing to what you are allowing here with the brain-washed American "bushified" soldiers is the destruction of Life. So who is the guilty ones here. We are honouring and having reverence for life.

Judge: Yes, but I understood from other records before me, that Jain is not the biological father of this child Galatea.

Phi-Lips: Yes, this is true, he is not the dna father, but does it really matter. The true father of this child, Jain, is the Love aspect that nourishes this child's soul. Galatea is her own entity, a fresh beginning because her earthly father opted out, blatantly abandoned the family out of fear, not love, therefore she could not take his family name, nor take my surname, and I registered her as "Father Unknown" and with an invented surname, this is the Truth, she has no roots, no lineage name...so I just surnamed her Prana.

And to conclude and end this debacle, I have no interest in your War Games, it against my Soul, I call on Jain to assist me out of this Chamber, and should you desire to continue this Farce, I ask you kindly to do it without my Presence, and without Jain 108.

In Love and In Peace,

Sephira Phi-Lips

Jain 108

Galatea Prana

And on behalf of all the **Unborns**,

I take leave.

(typed up on Nov 27th, 2007, Mullumbimby Creek)

 nb: Everything in this account, is true, that is, not fictional, all names are partially real or slightly changed, not invented!

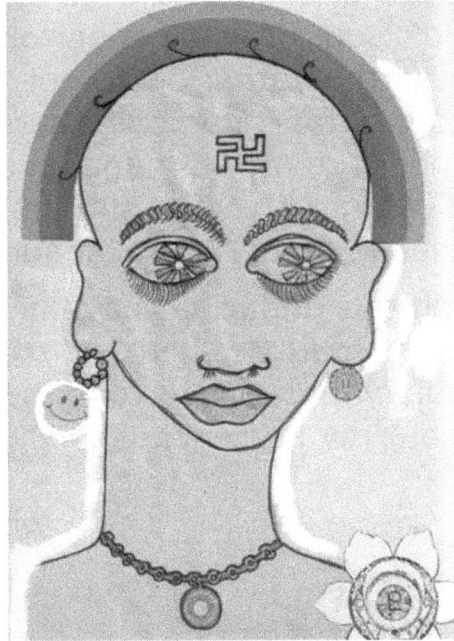

Art by Jain 1982 "La Boheme" drawn in the Torres Strait Islands
where I spent 2 years with native medicine men
learning by "transmission", without words, the living maths of nature.

For irene. EveryThing is for The Cosmic
Beloved

drink of me

(Art by Jain, 1986 "Cosmic Beloved)

The NEXT TWO BOOKS On PHI in this SERIES:

THE BOOK OF PHI, Vol 4 Sub-titled: PHI CODE 1

* Why 108? (= 40 pp)
* 108 PC1 as 4x6 and 3x8 Matrices (= 10 pp)
* Phi Code 1 (PC1) Mystic Cogged Wheels (= 46 pp)
* Jain's Phi-Pi Connection (= 46 pp)
   - **Part 1**: Geometrical Derivation of "1 + Root 5 divided by 2"
   - **Part 3**: **Jain's True Value of Pi = 3.144...**
   - **Part 4**: Where else does Phi and Pi Exist?
* Jain's Phi-Prime Number Connection (= 42 pp)

CHAPTER TITLES & CHAPTER DESCRIPTIONS
& PAGE COUNTS for "THE BOOK OF PHI, Volume 4"

* **Why 108?** (40 pages). An excellent numerical account of the fascination for the sacred number 108 throughout history and in various cultures.
* **108 Phi Code 1 as 4x6 and 3x8 Matrices** (10 pages). Gives various views or perceptions of the data relevant to the digital compression of the Fibonacci Sequence generating the infinitely repeating 24 Pattern.
* **Phi Code 1 Mystic Cogged Wheels** (46 pages). Plugs in the 24 Repeating Pattern into circular rings that reveal another whole realm of creative possibilities that would inspire scientists to make practical application of the 108 Code.
* **Jain's Phi-Pi Connection** (46 pages). For the first time in print, this rare knowledge of **Jain's True Value of Pi = 3.144...** is revealed, being based on the Square Root of Phi 1.272... in contrast to the erroneous and orthodox view of pi = 3.141...
* **Jain's Phi-Prime Number Connection** (42 pages). Many scholars have asked if there exists a connection between Prime Numbers and the Phi Ratio, and it is revealed here.

THE BOOK OF PHI, Vol 5 Sub-titled: PHI CODE 2

* Powers Of Phi Proved With 77 DP (= 46 pp)
* Derivation Of Enneagram From PC2 (= 31 pp)
* 108 Phi Code 2 (PC 2) as 4x6 and 3x8 Matrices (= 7 pp)
* PC2 Mystic Cogged Wheels   (= 37 pp)
* The 13 Code: Rings of 13 (= 28 pp)
* The 24 Cell And The Phi Code (= 4 pp)

CHAPTER TITLES & CHAPTER DESCRIPTIONS
& PAGE COUNTS for "THE BOOK OF PHI, Volume 4"

* Powers Of Phi Proved With 77 DP (46 pages). When the Powers of Phi are expressed on the electronic calculator using only 9 decimal places, the Pattern breaks down, but when 77 decimal places are used, the magic is revealed. A true tribute to the wonders of Technology. References to the Lucas Series.

* Derivation Of The Enneagram From Phi Code 2. (31 pages) Shows how the 9 Pointed Star Enneagram is generated by only using the geometrical constructions attained from the 108 Code of the Powers of Phi.

* 108 Phi Code 2 as 4x6 and 3x8 Matrices (7 pages). Originally we portrayed the 24 Repeating Pattern as a 2 x 12 rectangular array, but we can also express the same 24 sequence in various creative 4x6 and 3x8 Grids.

* Phi Code 2 (aka PC2) Mystic Cogged Wheels (37 pages). Extraordinary circular motifs designed totally from this 24 Repeating Pattern of the Powers of Phi. Great resource material for artists, physicists, scientists, and all futuristic designers.

* The 13 Code: Rings of 13 (28 pages). Uses digital compression to X-Ray the well known 13 Times Multiplication Table written circularly and revealing the anointed frequencies of 666 and 108

* The 24 Cell And The Phi Code (4 pages). Ultimately we must ask what shape in the Universe embodies this 24ness, apart from the Star Tetrahedron that has 24 faces and 24 edges. I would like to conclude this introduction by including a hand drawn diagram of the 5 arms of Ancient Knowledge that I specifically teach, it is designed in the Phi Rectangle format, the text surrounds a pentacle with the all-seeing eye, and the subject of the Phi Codes 1 & 2 are shown as both legs of the star.

Jain 108 Mathemagics
Sacred Geometry Mystery School

- Level 1. Beginners Course — 5 Days of Remembering
- Level 2. Advanced Course — 5 Days of Awakening
- Level 3. Teacher Training — 5 Days of Mastery

Art of Number

Level 1: • Digital Compression of the Multiplication Table revealing Atomic Structure of Rutile + Platinum Crystal

Level 2: • Prime Numbers 24 Pattern + 4th Dim Templar Cross
• Binary Code   origin of VW Symbol

Rapid Mental Calculation

Level 1:
• Multiplicat...
• Magic Fingers
Level 2:
• One Line Division
• Square + Cube Roots
• Fractions

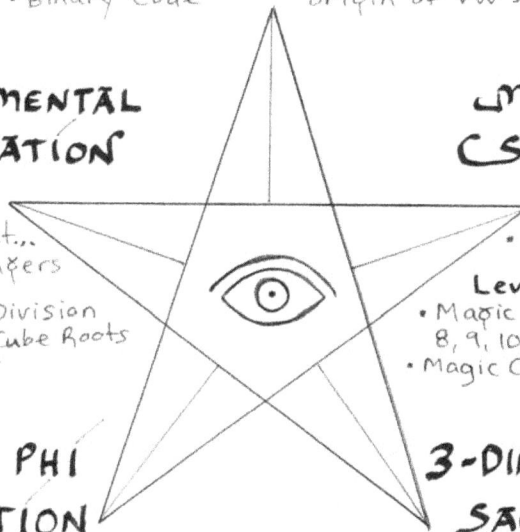

Magic Squares

Level 1:
• Magic Sqs. of 3, 4, 5, 6, 7
Level 2:
• Magic Squares of 8, 9, 10, 11, 12, 16
• Magic Cubes + Stars

Divine Phi Proportion (108 Codes)

Level 1: • Phi Spiral
• Fibonacci Numbers
• 108 Phi Code 1
Level 2: • Pentagram Constr.
• 108 Phi Code 2
• Mystical Squaring of Circle
• Vesica Pisces
(the Mother of all Form)

3-Dimensional Sacred Geometry

Level 1:
• the 5 Platonic Solids
• Fold Up Net of Dodecahedron
Level 2:
• the 13 Archimedean Solids
• Cuboctahedron perfected
• Truncated Icosahedron (Soccer Ball Shape).

Jain F.R.E.E.D.O.M.S Non Profit Org...
trading as: JAIN MATHEMAGICS
www.jainmathemagics.com

SOME ENVIRONMENTAL OCCURENCES
of
108 and or 1.618

Here follow some examples where the anointed numbers like 108 or 1.618 appear.

The following occurrences are natural, and have not been faked or reproduced to make them sensational for purposes of this book. They are real events in this E8 paradigm:

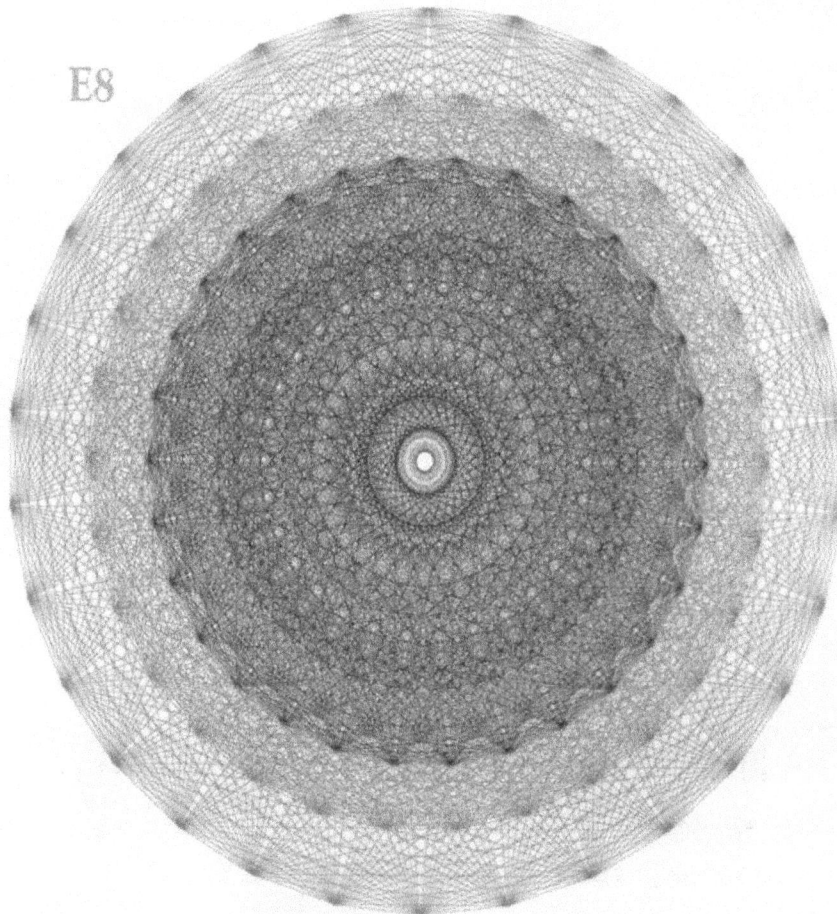

(E8 Field: Taken from Dan Winter's website,
may originate from another author Nassim Harrimein of Hawaii)

INCIDENT NUMBER 1

March, 2009: The photo below shows my arrival at the Amsterdam Schiphol International Airport bay, where we are being ferried by bus to anther connecting flight, and the number just happens to be "108". The first photo shows a long distance shot, and then the second photo shows a close up of the number 108

INCIDENT NUMBER 2

Dec 2008

My partner Sephira was at a mall, the Pines Shopping Centre in the Gold Coast, seeking to buy some books for her 18 month daughter Galatea. To my surprise, she showed me this adorable book: "**THE 108th SHEEP**" by **AYANO IMAI** copyright 2006, by www.koalabooks.com.au

The **back cover blurb**:

Nimitz can't sleep a wink, and so she decides to count sheep to help her fall asleep. One by one, the sheep jump over her bed and Nimitz counts them until she discovers that the 108th sheep is stuck. Luckily, Nimitz has a plan to save the night.

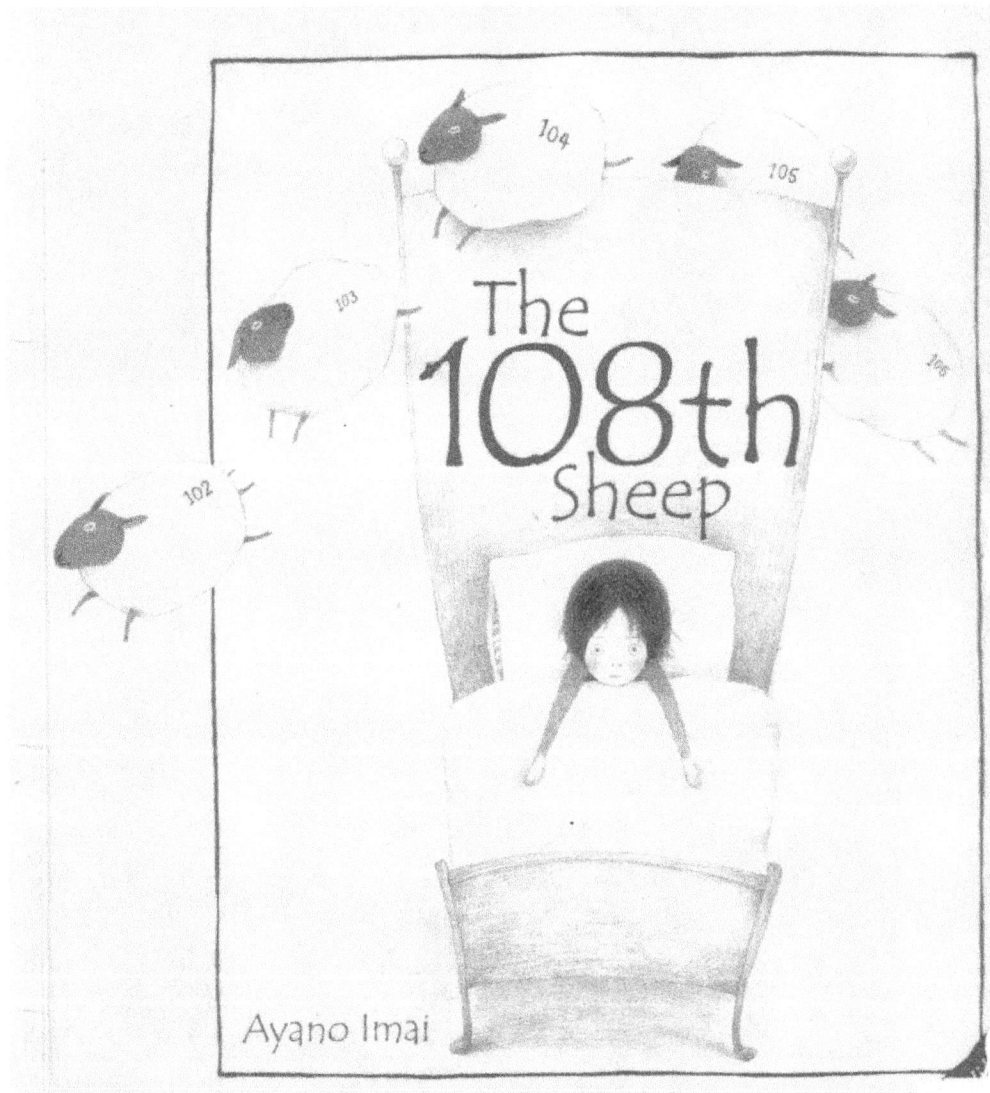

INCIDENT NUMBER 3

January 2009: Another incident, I am driving up the coast, to see my two daughters Mingkah and Aysha Sun, and stop half way to refuel.

What do you think the cost was to fill it up?

It cost exactly **$61.80 cents**, a harmonic of the divine proportion, **1:1.618...**

The receipt is very poor quality and did not scan very well, but it is included to show that these **synchronicities** happen all the time, when you are passionate about a certain topic, you magnetize to you that frequency.

June 2009 Here is another fuel receipt from Tugun, Gold Coast, Queensland, Australia, showing that when I filled up with fuel, the total amount just happened to be **$61.80 cents**. It just keeps happening.

INCIDENT NUMBER 4a

Feb 2009 Time to renew my driving license, so off I go to the RTA (Road Traffic Authority) in Tweed Heads, far north NSW. These days you are required to collect a ticket with a number which puts you in a queue, and you wait until your number flashes up on the screen, so you can proceed to the counter. Have a guess what my ticket number was:

That's right: **Number 108**. And just to prove it, I kept a copy of that ticket:

RTA
Roads and Traffic Authority
Tweed Heads
Welcome
108
Licence Transactions
Please take a seat and watch
for your number to be displayed
2:24pm 3/2 2009   1536

Whilst I was standing in the queue, and receiving this ticket, you can see the time at the bottom line being recorded as 2:24pm. Thus prior to this, when I sat in my vehicle ready to drive off of the RTA, I happened to look at the time, written in digital language, and of course, it was **1:08pm**!

INCIDENT NUMBER4b

Not only was the RTA ticket number "**108**", the actual price that I paid for the license renewal also was $108 precisely. To verify this, I have also scanned my actual Westpac Mastercard bank statement, it is appears in the following diagram, shown as underlined:

Westpac

'No Annual Fee' MasterCard Statement

Account Number	5163 2010 3010 3630
Due Date	12 Mar 09
Closing Balance	627.24 -
Minimum Payment Due	0.00
Amount Paid (Details on the reverse)	$

ılıılıılıılıılıʺıllıııılıılıllllllıʺʺʺʺʺıllıllllıllı   018

JAIN JAIN
777 LEFT BANK RD
MULLUMBIMBY CREEK NSW 2482

+5163201030103630+

....

(Cut along this dotted line)

For enquiries and other services please call the Cards Customer Service Call Centre on **1300 651 089** from anywhere in Australia, 24 hours a day, 7 days a week or visit our internet site at **westpac.com.au.** For payment methods please see reverse.

Any statement entries for purchases or cash advances/withdrawals made in a foreign currency include the following three components: (1) the foreign currency transaction amount, converted into Australian dollars by the applicable credit card scheme; (2) any fee that may be charged by the applicable credit card scheme to Westpac on foreign currency transactions; and (3) a foreign currency transaction fee charged to you by Westpac, being 2.0% of the Australian dollar transaction amount.

Cardholder Name		Account Number	Customer Number	Credit Limit	Available Credit
JAIN JAIN		5163 2010 3010 3630	08972883	5,000	5,627.24

No. of Days in Statement Cycle	Statement From	Statement To	Minimum Payment Due	Due Date	Opening Balance
33	14 Jan 09	15 Feb 09	0.00	12 Mar 09	28.29 -

Date of Transaction	Description	Debits	Credits (-)
14 Jan 09	TICKETMASTER7 SOUTHBANK AUS	125.50	
20 Jan 09	PAYMENT - CASH THANK YOU		200.00 -
21 Jan 09	VISTAPR*VISTAPRINT.CA 800-961-2075 GBR	124.63	
24 Jan 09	ROBINS KITCHEN BURLEIGH HEAD AUS	110.98	
29 Jan 09	PAYMENT - CASH THANK YOU		700.00 -
01 Feb 09	AMPOL 28125 BRUNSWICK HEA AUS	61.80	
03 Feb 09	395 RTA TWEED HEADS TWEED HEADS S AUS	108.00	
04 Feb 09	PAYMENT - CASH THANK YOU		400.00 -
06 Feb 09	BP CHINDERAH 2424 CHINDERAH AUS	60.36	
08 Feb 09	NONNAS ITALIAN CUISINE BIGGERA WATER AUS	34.50	
08 Feb 09	VICTORIA STATION BIGGERA WATER AUS	250.90	
08 Feb 09	TAROCASH BIGGERA WATER AUS	179.80	
08 Feb 09	FAMOUS FOOTWEAR BIGGERA WATER AUS	104.97	
08 Feb 09	ESSENTIAL MAN BIGGERA WATER AUS	34.00	
08 Feb 09	ED HARRY MENSWEAR BIGGERA WTRS AUS	104.95	
11 Feb 09	PAYMENT - CASH THANK YOU		150.00 -
11 Feb 09	PAYMENT - CASH THANK YOU		450.00 -
15 Feb 09	INTEREST CHARGES - PURCH	0.66	

INCIDENT NUMBER 5

I am driving along, towards North Bondi Beach, and happen to stop on the side of the kerb, just to check the roadmap, and have a guess what house number I perchanced to view:

INCIDENT NUMBER 6

Driving back home from Sydney to Mullumbimby along the
highway, with my cosmic beloveds Sephira and Galatea, what were
the distances listed on the large green road sign:

INCIDENT NUMBER 7
108 OCCURENCES   cont....

Here is another incident involving the number 108 again, no picture with this one, just an email verification.

I sent an email to Lila of London to organize some contract details for an upcoming seminar, and when it was sent, I noticed that the time at the bottom of the email was **1:08 PM**

Lila Hammond wrote:

Hi Jain

I am glad the dates work for you too. I'll start advertising and will also send you my additions to my contract soon.
I know you said you had information on your site for the work/teaching you do and I will take from it what I need.
Could you also send me a write up for this particular course as well please?
And yes, we can speak on Skype, maybe this coming weekend?
With love
Lila
http://www.thisisenergy.com
--- On Thu, 11/6/09, Jain 108 <jain@jainmathemagics.com> wrote:

From: Jain 108 <jain@jainmathemagics.com>
Subject: Re: Contract - UK Workshop
To: "Lila Hammond" <lilahpav@yahoo.com>
Date: Thursday, 11 June, 2009, **1:08 PM**

EARTH EARTH

(or EARTHHEART) LEXIGRAM

(since the anagram of EARTH is HEART).

Using only the Letters: A - A – E – E – H – H – R – R - T - T

NUMBER OF LETTERS
(listed in alphabetical order)

1	2	3	4	5	6	7
A	AH	AHA	HARE	EARTH	HATETH	HEARETH
	AT	ARE	HATE	EATER	HEARER	THEATRE
	HA	ART	HEAR	HATER	HEARTH	
	HE	EAR	HEAT	HEART	HEATER	
	RE	HAT	HERE	HEATH		
		HER	RARE	TEETH		
		RAT	RATE	THERE		
		TAR	REAR	THREE		
		TEA	TARE	TREAT		
		TEE	TART			
			TEAR			
			TEAT			
			THAT			
			THEE			
			TREE			

Here is the lexigram poem based on the above words of
EARTH EARTH:

title: **HEAR THE EARTH HEART**

AH
HEAR THEE
HEAR THE EAR
HEAR HER ART
HEAR THE EARTH
HER RARE EARTH
HEAR THE HEART RATE
HERE THE EARTH HEARER
HEAR THEE THE EARTH HERE
HEAR THE HEART HERE
HEARETH THE HEART
HEAR THE EAR HERE
HEAR THEE HERE
HEAR THE TREE
EARTHHEART
A TREE
HERE
AHA
A TEAR
TEETH AT TEAT
A EARTH HATER
A HEART HATER
A HEATH HATER
HERE AT EARTH
THE HEART HATE
HATE THE HATER
HE HATE HER HEAT
HE TEARETH HER HEART
THE HATE TEAR AT THE HATER
THE HATE TEAR AT THE EARTH
HE HATE HER HEART HER TEETH
THE HATE TEARETH AT THE EARTH
HE HATETH THE RAT THAT ATE THE TEA TART TREAT
THE HEART TEAR AT THE RAT HATER
A TEAR AT THE HEART THEATRE
HER HEART TEAR AT THE HATE
HE HER HATE TAR THE EARTH
HE RATE THE EARTH TAR
A HATER AT THE EARTH

TREAT EARTH HEARTH
HATE ATE THE EARTH
A TREE HATER TEAR
HATE TREE EATER
TREAT HER HEART
HEAR HER ART
EARTH ART
THEATRE
HEAR
THE EAR
HERE THERE
HEAR THE EARTH HEART
HERE THE EARTH HEART THEATRE
HEAR THEE HER HEART
EARTH HEART ART
RARE EARTH

by
JAIN 108
mullumbimby creek, 2008

(Art by Jain, 2009, "Reflecting Upon Stillness in the Now")

PRODUCT OF THE PHIBONACCI NUMBERS

JAIN 108
18-8-2009 Mullumbimby Creek.

Thanks to the **Ccalc** Calculator, free downloadable online Calculator that allows accuracy up to **80 Decimal Places**, without which I could not have recorded the following data.

Aim is to determine that the Digital Sums of the products of the **Phi**bonacci (aka Fibonacci) Numbers have a recognizable repetition or recursion after 24 products, since there is a hidden 24 repeating pattern in the two Phi Code 108 Sequences.
Or do all the Digital Sums reduce to 9, after the 7th product, regardless of the 24 Repeating Pattern?

Legend:

pFn1	= the product of the first 2 Phibonacci Numbers ie: 1x1
pFn4	= the product of the first 5 Phibonacci Numbers ie: 1x1x2x3x5
DP	= Decimal Places.
Ans	= Answer
*****	= Multiplication
>	= a space separating text from numbers, does not mean the symbol for "greater than".

pFn1 > 1*1
ans = 1
Digital Sum = **1**

pFn2 > 1*1*2
ans = 2
Digital Sum = **2**

pFn3 > 1*1*2*3
ans = 6
Digital Sum = **6**

pFn4 > 1*1*2*3*5
ans = 30
Digital Sum = **3**

pFn5 > 1*1*2*3*5*8
ans = 240
Digital Sum = **6**

pFn6 > 1*1*2*3*5*8*13
ans = 3,120
Digital Sum = **6**

pFn7 > 1*1*2*3*5*8*13*21
ans = 65,520
Digital Sum = 18 = **9**

pFn8 > 1*1*2*3*5*8*13*21*34
ans = 2,227,680
Digital Sum = 27 = **9**

pFn9 > 1*1*2*3*5*8*13*21*34*55
ans = 122,522,400
Digital Sum = 18 = **9**

pFn10 > 1*1*2*3*5*8*13*21*34*55*89
ans = 10,904,493,600
Digital Sum = 36 = **9**

pFn11 > 1*1*2*3*5*8*13*21*34*55*89*144
ans = 1,570,247,078,400
Digital Sum = 45 = **9**

pFn12 > 1*1*2*3*5*8*13*21*34*55*89*144*233
ans = 365,867,569,267,200
Digital Sum = 72 = **9**

pFn13 > 1*1*2*3*5*8*13*21*34*55*89*144*233*377
ans = 137,932,073,613,734,400
Digital Sum = 63 = **9**

pFn14 > 1*1*2*3*5*8*13*21*34*55*89*144*233*377*610
ans = 84,138,564,904,377,984,000
Digital Sum = 90 = **9**

pFn15 > 1*1*2*3*5*8*13*21*34*55*89*144*233*377*610*987
ans = 83,044,763,560,621,070,208,000
Digital Sum = 72 = **9**

pFn16 >
1*1*2*3*5*8*13*21*34*55*89*144*233*377*610*987*1597
ans = 132,622,487,406,311,849,122,176,000
Digital Sum = 90 = **9**

pFn17 >
1*1*2*3*5*8*13*21*34*55*89*144*233*377*610*987*1597*
2584
ans = 342,696,507,457,909,818,131,702,784,000
Digital Sum = 126 = **9**

pFn18 >
1*1*2*3*5*8*13*21*34*55*89*144*233*377*610*987*1597*25
84*4181
ans = 1,432,814,097,681,520,949,608,649,339,904,000

pFn19 >
1*1*2*3*5*8*13*21*34*55*89*144*233*377*610*987*1597*25
84*4181*6765
ans = 9,692,987,370,815,489,224,102,512,784,450,560,000

pFn20 >

1*1*2*3*5*8*13*21*34*55*89*144*233*377*610*987*1597*25
84*4181*6765*10946

ans =

106,099,439,760,946,345,047,026,104,938,595,829,760,000

pFn21 >

1*1*2*3*5*8*13*21*34*55*89*144*233*377*610*987*1597*25
84*4181*6765*10946*17711

ans =

1,879,127,177,606,120,717,127,879,344,567,470,740,879,360,00
0

pFn22 >

1*1*2*3*5*8*13*21*34*55*89*144*233*377*610*987*1597*25
84*4181*6765*10946*17711*28657

ans =

53,850,147,528,658,601,390,733,638,377,270,009,021,379,819,5
20,000

pFn23 >

1*1*2*3*5*8*13*21*34*55*89*144*233*377*610*987*1597*25
84*4181*6765*10946*17711*28657*46368

ans =

2,496,923,640,608,842,029,285,537,344,277,255,778,303,339,47
1,503,360,000

Digital Sum = 225 = **9**

pFn24 >

1*1*2*3*5*8*13*21*34*55*89*144*233*377*610*987*1597*25
84*4181*6765*10946*17711*28657*46368*75025

ans =

187,331,696,136,678,373,247,147,439,254,401,114,767,208,043,
849,539,584,000,000

Digital Sum = 198 = **9**

pFn25 >
1*1*2*3*5*8*13*21*34*55*89*144*233*377*610*987*1597*25
84*4181*6765*10946*17711*28657*46368*75025*121393
ans =
22,740,756,589,119,797,763,590,969,093,409,514,524,935,686,0
67,027,158,720,512,000,000
Digital Sum = 279 = **9**

pFn26 >
1*1*2*3*5*8*13*21*34*55*89*144*233*377*610*987*1597*25
84*4181*6765*10946*17711*28657*46368*75025*121393*1964
18
ans =
4,466,693,927,721,732,437,129,010,967,389,310,023,958,817,58
5,913,340,461,565,526,016,000,000

pFn27 >
1*1*2*3*5*8*13*21*34*55*89*144*233*377*610*987*1597*25
84*4181*6765*10946*17711*28657*46368*75025*121393*1964
18*317811
ans =
1,419,564,463,863,171,507,576,408,104,556,964,008,024,375,77
5,796,704,645,430,601,388,670,976,000,000

pFn28 >
1*1*2*3*5*8*13*21*34*55*89*144*233*377*610*987*1597*25
84*4181*6765*10946*17711*28657*46368*75025*121393*1964
18*317811*514229
ans =
7.2998121468789482116950876319822304488236673081216363
31151327214948873175040000**e80**

pFn29 >
1*1*2*3*5*8*13*21*34*55*89*144*233*377*610*987*1597*25
84*4181*6765*10946*17711*28657*46368*75025*121393*1964
18*317811*514229*832040
ans =
6.0737356986891600700587807133144950226392441470495262
929711502959260604365602816**e86**

pFn30 >

1*1*2*3*5*8*13*21*34*55*89*144*233*377*610*987*1597*25
84*4181*6765*10946*17711*28657*46368*75025*121393*1964
18*317811*514229*832040*1346269

ans =
8.17688208533855683835796465213319189963351257860421871
29119775377460814578675737**e92**

CONCLUSION:

Assuming all the sums add to **9**, (from pFn8 to pFn30) we can
temporarily state that the Sequence of Digital Sums or Digital
Compressions of the Product of the Phibonacci Numbers is:
= 1, 2, 6, 3, 6, 6, 9, 9, 9, 9, 9, 9, 9, 9, 9, 9, 9, 9, 9, 9, 9, etc

= 1, 2, 6, 3, 6, 6, 9 repeater...

= **1, 2, 6, 3, 6, 6, 9**
(in mathematical language, the repetition of the 9 is usually
indicated with a small dot on top of the 9, here it is shown as
underlined).

It's interesting that the 6 numbers above: (1, 2, 6, 3, 6, 6,) add up
to 24 which is akin to the hidden 24 repeating pattern in the two
Phi Code patterns. Then all numbers repeat as 9.

To confirm this as an official paper, I still need to plug this data into
a larger computer that offers more than 80 DP. We notice in the
data that pFn28 breaks down, and is not sufficient to hold the true
answer with 80 DP'

It would also be of use to find or create some computer software
that can quickly digital compress larger numbers with hundreds of
digits, as I have been doing this manually, which is time-
consuming.

This is an ongoing and evolving topic, and I welcome any new
information on this subject.

COSMIC MATHS LOGOS:

As a tribute to all Mathematics Teachers, to honour the immense work of them just holding and sharing the knowledge, by teaching the subject year after year, same material year after year, I must thank you for keeping the torch alive, 'cos all mathematics is beautiful, it just keeps getting better in the sense that it keeps evolving it's language, like a living sculpture...
Here are 2 cosmic logos found on maths books:

I am always delighted when I come across old and dusty mathematic books whose front cover or the logo of the publishers is of high sacred geometrical value, verifying that Mathematics, indeed, is a Star Language. Any maths book, no matter how dry or cold, still has immense value. This is expressed in the words of Bertrand Russell:
"Mathematics, rightly viewed, possesses...supreme beauty, cold and austere, like that of sculpture, without appeal to any part of our weaker nature."
The books, upon opening them, may scare the average reader, but their logos are mathematical gems or plums depicting 2 phi spirals in the fractal heart form, and the shadow of the dodecahedron!

1 – dodecagonal shadow (ie: dec view or 10 sided) of the dodecahedron, shown on front cover of this maths book:

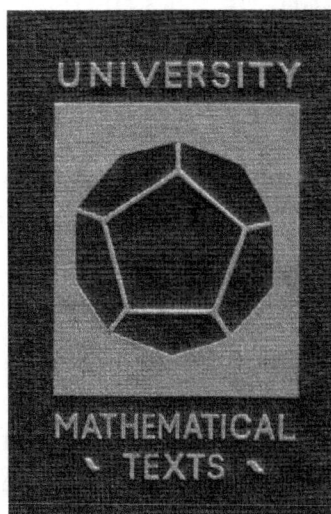

2 – Shown below are 2 phi spirals shown as the logo for Carslaw Publications, who published a maths books of undergraduate lecture notes, called "Combinatorics. An Introduction" by K.H. Wehrhahn in 1990.

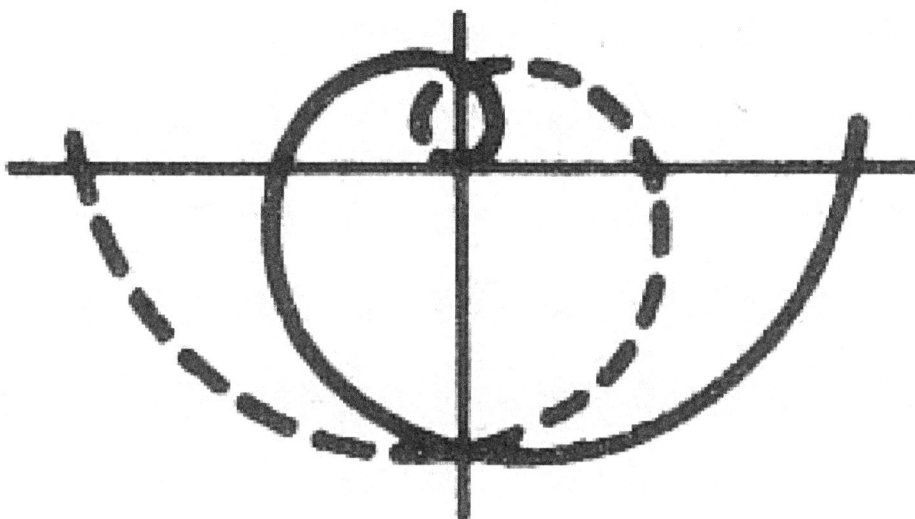

("Introduction To The Theory Of Finite Groups"
by Walter Ledermann, Ph.D, D.Sc. Published by Oliver and Boyd, 1964).

aPPENDIX 6

SQUARE ROOT OF 5

```
  2.23606797749978969640917366873127623544061835961152572427089724541052
□ +092563780489941441440837878227496950817615077378350425326772444707386
□ +358636012153345270886677817319187916581127664532263985658053576135041
□ +753378500342339241406444208643253909725259262722887629951740244068161
□ +177590890949849237139072972889848208864154268989409913169357701974867
□ +888442508975413295618317692149997742480153043411503595766833251249881
□ +517813940800056242085524354223555610630634282023409333198293395974635
□ +227120134174961420263590473788550438968706113566004575713995659556695
□ +691756457822195250006053923123400500928676487552972205676625366607448
□ +58535052623306784946334222423176372770266324076801044431582573350589
□ +309813622634319868647194698997018081895242644596203452214119223291259
□ +819632581110417049580704812040345599494350685555185557251238864165501
□ +02643631257102444961878942468290340447471611545572320173767659046091
□ +852957560357798439805415538077906439363972302875606299948221385217734
□ +859245351512104634555504070722787242153477875291121212118433178933519
□ +103800801111817900459061884624964710424424830888012940681131469595327
□ +944789899983169157746079246180750067987712420484738050277360829155991
□ +396244891494356068346252906440832794464268088898974604630835353787504
□ +206137475760688340187908819255911797357446419024853787114619409019191
□ +368803511039763843604128105811037869895185201469704564202176389289088
□ +444637782638589379244004602887540539846015606170522361509038577541004
□ +219368498725427185037521555769331672300477826986666244621067846427248
□ +638527457821341006798564530527112418059597284945519545131017230975087
□ +149652943628290254001204778032415546448998870617799819003360656224388
□ +640963928775351726629597143822795630795614952301544423501653891727864
□ +091304197939711135628213936745768117492206756210888781887367167162762
□ +262337987711153950968298289068301825908140100389550972326150845283458
□ +789360734639611723667836657198260792144028911900899558424152249571291
□ +832321674118997572013940378819772801528872341866834541838286730027431
□ +532022960762861252476102864234696302011180269122023601581012762843054
□ +186171761857514069010156162909176398126722596559628234906785462416185
□ +794558444265961285893756485497480349011081355751416647462195183023552
□ +595688656949581635303619557453683223526500772242258287366875340470074
□ +223266145173976651742067264447621961802422039798353682983502466268030
□ +546768767446900186957209958589198316440251620919646185105744248274087
□ +229820410943710992236175285315302212109176295120886356959716907946257
□ +260325089752229704043412880822332153390119551566514079022175646165421
□ +295787804223138207855367690772666643131659319546206872064645091487274
□ +408248812817765347516867907359186246442687464199149977893991312947201
□ +459199967825762063948526250359428286402462255910378955634538283178235
□ +598391296251160036910131265905719718200181724360595512757851998329989
□ +285638604458710469334951865390330842804218272603638944541578024417457
□ +472341469729999631251094562274695974331390549780162887681065496756275
□ +649338348884592698294163140147050914141795453509386876452390937230662
□ +419067158476029218547020420238380436721350194617915057915493628459086
```

taken from: http://www.perlmonks.org/?node_id=602366

bIBLIOGRAPHY

~ **The Book Of Phi, Volume 1**, sub-titled: **The Living Mathematics Of Nature**, by **Jain**, self-published 2002

~ **The Book Of Phi, Volume 2**, sub-titled: **In The Next Dimension**, by **Jain**, self-published 2003.

~ "**Numbers Of Light**" by **Jason O'Hara**, 2007 London
http://www.twentyfourthmystery.net
This book starts off with the 24 Repeating Pattern and evolves to the 4th Dimensional Maltese Cross and links to atomic structure.

~ "**The Language Of Pattern**", by 4 students of Keith Critchlow: **Keith Albarn**, Jenny Miall Smith, Stanford Steele, Dinah Walker who first showed in printed form in 1974 the 24 pattern though without any detail. Published by Thames and Hudson, London.

~ **www.phinest.com** by **Gary Meisner -usa** who published online the 24 repeating pattern, acknowledging both Tuberville and Jain. Also shows excellent examples on the 108 decimal for the curious reciprocal of the number 109.

~ "**A Glimmer of Light from the Eye of a Giant** "by **Joseph Turbeville**, independently cracked the 24 Repeating pattern.
On 7-9-06 he emailed me:
Dear Sir Jain, I have just recently discovered your most impressive web site, and was particularly interested in the Lost Secrets of the Phi Code that you write about. I too have also been intensely involved in quite similar research for several years, and I'm sure you have not been made aware of it, or you would have certainly made reference to my work in the area that matches yours. I do hope you will look into my research as it deals with the **distillation** (i.e. **mathematical reduction**) and tabularization of the Fibonacci series, **Lucas Series** and sets of sequential numbers. My tables have been called the **Glimmer Tables** or some have referred to them as the Turbeville Tables. I do believe that after you review my research findings of the ancient codes of nature you will be as

excited with my discoveries as I was. My first book was, 'A Glimmer of Light from the Eye of a Giant' -- **www.trafford.com** - in 2000. They have sites for writers. My personal website is **www.eyeofagiant.com** -- and I have articles posted at various sites on the www that you can just "Google" my name to find them. I hope this will be a mutually beneficial contact.

Respectfully submitted Joseph Turbeville.

Art by Jain, 2004 "Learning not via physical books but by accessing the Akashic Records within"

eXTRA nOTES aND cOMMENTS:

Page 1: The diagram shows The Art of Jain with the Phi Code 108 string or necklace of 24 infinitely repeating digits. But, there is another sequence of 18 digits curving around the crown of the person, made up of 3 lots of 6 digit codes: **1-2-4-8-7-5**. This periodicity of 6 digits is the well known Digital Compression of the **Binary Sequence**, discussed in Chapter 3 called: "Binary Code Vs Phi Code" beginning on page 84 of this book.

This symbolically links the Phi Codes with the Binary Codes, which is explained well in Chapter 3, by combining both sequences on this one diagram. There is no duality. All is interconnected. The 108 Phi Code is no more spiritual than the language of computers symbolized by the Binary Code 1-2-4-8-7-5.

15/07/2005 07:43 am

Art By Jain, 1984, "Divine Being"

i N D E X

LEGEND:

~ Numbers in **Bold**, represent reference to an image.

~ "aka" means "Also Known As".

~ "Ch" means "Chapter".

A

B

CUP (WITHIN THE CUP): 85,

FRACTALITY: 11, 13, 50, 64, 91, 102, 114, 136, 194,
FRANCIS OF ASSISSI: 3,
FRANKENSTEIN: 148,
FREEMASONRY: 101,

G

GALACTIC MATHEMATICS: 29,
GALATEA: 140-150, **151**, **152**, **153**, 157, 168-170, 177, (Original Person that this Entity is based upon is Starr Prana aged 2 years, daughter of Gabrielle Phillips aka Sephira, used with her mother's permission).
GARZIOTTI, Adriano: **114** (Polyhedra),
GAYATRI MANTRA: 29, 162-163-164,
GENE-SPLICING: 150,
GLOBALIZATION: 58, 63-64, 103, 147,
GOAT HORN: **22**,
GOD: 74, 147,
GODLEN MEAN SPIRAL: see Phi Spiral.
GRAPHICAL DERIVATION OF PHI: 76,
GRATITUDE: 32,
GYORGY DOCZI: 19

H

HAHA-AHA: **65**,
HANLEY, Prof Bill: 140-150,
HARMONIC STAIRWAY: 77-78,
HARMONICS: **73-74**, 77, 129 (of Light), 134,
HEART: 33, 50, 129 (Fractal), 136, 148 (Attack), 151 (Attack), 194,
HEMACHANDRA-GOPAL: 39,
HERMETIC: 12,
HEX VIEW (Shadow): 112,
HITLER: 104,
HOMEOPATHICS: 129 (Base 10), 135 (Digital),
HYPER-SPACE: 153,
HYPOTHALAMUS: 69,

I

ICOSAHEDRON: **25**, **113**-114,
ILLERT, Chris, 116,
IMAGINEERING: 150,
IMBUED: 168,
IMPLOSION: 49, 74, 103,
INCUNABULUM: 12,
INDESIGN: 3,
INFINITE SERIES: **25**,
INFINITY: 49,
INFO-SYS (HYDERABAD): 164,

QUEEN OF ENGLAND: 61-62,
QUINTIN, Jonathan: 124,

R
RAPID MENTAL CALCULATION: see Vedic Mathematics.
RECIPROCALS: 80-82 (Reciprocal of 109), 129 (of 144 Bruce Cathie),
RECURSIVE: 27, 35, 38, 105,
RELATIONSHIFT: 149, 168,
REMEMBER: 32,
RETURN TO SENDER: 145,
REVERSALS: 80,
RHOMBIC DODECAHEDRON: 86, **115**,
RINGS OF 13: 173,
ROSE: 50,
ROSICRUCIAN CROSS: **98**-101,
RTA (ROAD TRAFFIC AUTHORITY): 179-180,
RUSSELL CROWE, 162,

S
SAAD, Peter: 140-150, **151**, 157-170,
SACERDOS LIBRI NATURAE: 83,
SACRED GEOMETRY: 10, 30, 56, 119,
SAHASRARA (aka Crown Chakra): 88, 102,
SCALE INVARIANCY: 25, 135,
SCHIPHOL AIRPORT AMSTERDAM: 176,
SEA-SHELLS: **22**, **23**,
SELF-SIMILAR: 27, 49, 91, 135,
SEPHIRA: 140-150, **151**, **152**, **153**, **154**, **155**, **156**, 157, 167-170,
    182, (Original Person that this Entity is based upon is Gabrielle Phillips,
    mother of Starr Prana aka Galatea, used with her permission).
SHAPE SHIFTING: 114, 150,
SHAPE STORES MEMORY: 30, 32, 88,
SHARING: 50, 116, 138, 148,
SHEEPLE: 61,
SHOALWATER BAY: 157,
SHRI 108: 163, see 108 in Number Section below.
SPHERES: **112**-113,
SPIRAL: 92 (fractal),
SPIRIT: 123,
SPIROS SPIROS: **118** (photo),
SPONE: 9, 63,
SQUARE ROOTS: 72(root of 5), 93 (root of 3), 95, 115(root of 2),
SQUARING OF THE CIRCLE: 49, 146,
STABBED THROUGH HEART: 167,
STAIRWAY TO HEAVEN: 116,
STAR LANGUAGE: 13, 63, 193,

**Art by Jain, 1999, Mural of Hathor, oil on canvas,
(a 4th Dimensional Venusian, cow-earred Being,
Sonic & Frequency Master)**

nUMBER index

**(NUMBER REFERENCES:
as part of JAIN'S DICTIONARY of NUMBERS
aka HARMONIC STAIRWAY)**

Legend:

... (3 Dots) after some numbers means that the decimal keeps on continuing.
: (Colon) means Proportion, as is 21:34, spoken as "the ratio of 21 to 34".

~ Numbers in **Bold**, represent reference to an image.

–1.618033988... — 72-73, 76,

0 — See Zero

.0069444444444... — 134 (Reciprocal of 144),

.00917431192660550458715596330275229357798165137614688990825688073394495412844036697247706422018348623 **853211 (Reciprocal of 109)** – 80-82 ,

.618...(Reciprocal of Phi = 1 divided by 1.618... or 1/ɸ) — 58, 71-72, 75-76,

1 — 49, 58, 88, 132,

1-2-6-3-6-6-9 repeater... — 192 (Digitally Compressed Sequence of Phibonacci Products)

1-2-4-8-7-5 — 1, 197 (Digitally Compressed Sequence of Binary Code Sequence),

1 : 3 : 3 : 1 — **43, 45**

1 : 5 : 10 : 10 — 39,

1.272...(Square Root Of Phi) — 49, 64-**65**,

1.414(Square root of 2) — 115,

1.618033988... — 16, **25**, 27, 93, 138, 175, 178,

1.732...(root of 3) — 93, 95,

2 — 88,

2.2360679... (Square Root of 5) — 72, 172-173,

3 — 91,

3.141592 (Traditional Pi) — 27, 103, 165, 172,

3.144... (JainPi) — 49, 165, 172, True Value Of Pi

4 — 88,

8 — 29, 88,

8:13 — 17,

9 — 28, **51**-52, 104-105, **109**, **131**, 187-192 (Phibonacci Products),

10 — **89**,

12 — 24, 28, **51**-52, 86, 113, **115**,

12:24:60 — 33, 35, 36, **37**, 38-49,

13 — 173 (Rings Of 13),

14 — **115**,

20 — 24, **25**,

21:34 — 18, 20, 31, 60, 90, **91**, 102, 163,

24 — 29, **30**, 31, 49, **58**, 60-61, 86, **117**, **118**, 172-173, 187-192
     (Phibonacci Products),

24 REPEATING PATTERN — 28, 78, 195,

32 — 123 (32°), 175 (E8 Mandala with 32 Points),

36 — 89, **90**,

46 — 97,

48 — 118,

60 — 35, 113,

61.80 — **178** ($61.80 cents),

69 — 149,

90° — 123, **127**,

108 — 29, 52, 53 (Sri 108), 77, 80, **83**, 130, 135, 157-170, 172-173, 175-
     178,
     179-183,

109 — 77, 79-82,

117 — 7, 77, 79,

216 — 87,

300 — 38,

360 — 150,

365 — 150,

366 — 48,

512 — 88,

666 — 63, 173,

695 — 134,

1,000 — 88,

2,012 — 47,

1,500 — 38-39,

15,000 — 38-39,

150,000 — 38-39,

10^{14} or **100,000,000,000,000** or **10 trillion** — 96 (cells in human
     body).

Jain lecturing at a 3 day Utopia Conference, Noosa Heads, Queensland. Oct 2009. As a guest speaker, I was teaching the audience of 100 people how to breathe into their Heart, called the Earthheart Meditation.
(Photo taken by Simone Matthews, one of the organizers).
Notice the wheel of 24 Phi Code numbers demonstrating the 12 Pairs of 9 that sum to 108.

THE BOOK
OF PHI
VOLUME 3

(SUB-TITLE)

The 108 Codes,
an Introduction

Phi Code 1   Phi Code 2

BY JAIN 108

Old Front Cover

THE BOOK OF PHI

VOLUME

1

THE LIVING
MATHEMATICS OF NATURE

Jain 108

PUBLISHED 2002

The Book Of Phi, volume 1, front cover

THE BOOK OF PHI

VOLUME

2

IN THE NEXT DIMENSION

Jain 108

PUBLISHED 2003

The Book Of Phi, volume 1, front cover

Art Work by Aurel Pumayana. See **www.luminaya.com**
The front cover of this book The Book of Phi, vol 3. utilizes some collaging of his original artwork.
Here, is a sample of his computer graphic art poster work: "Time Machine".

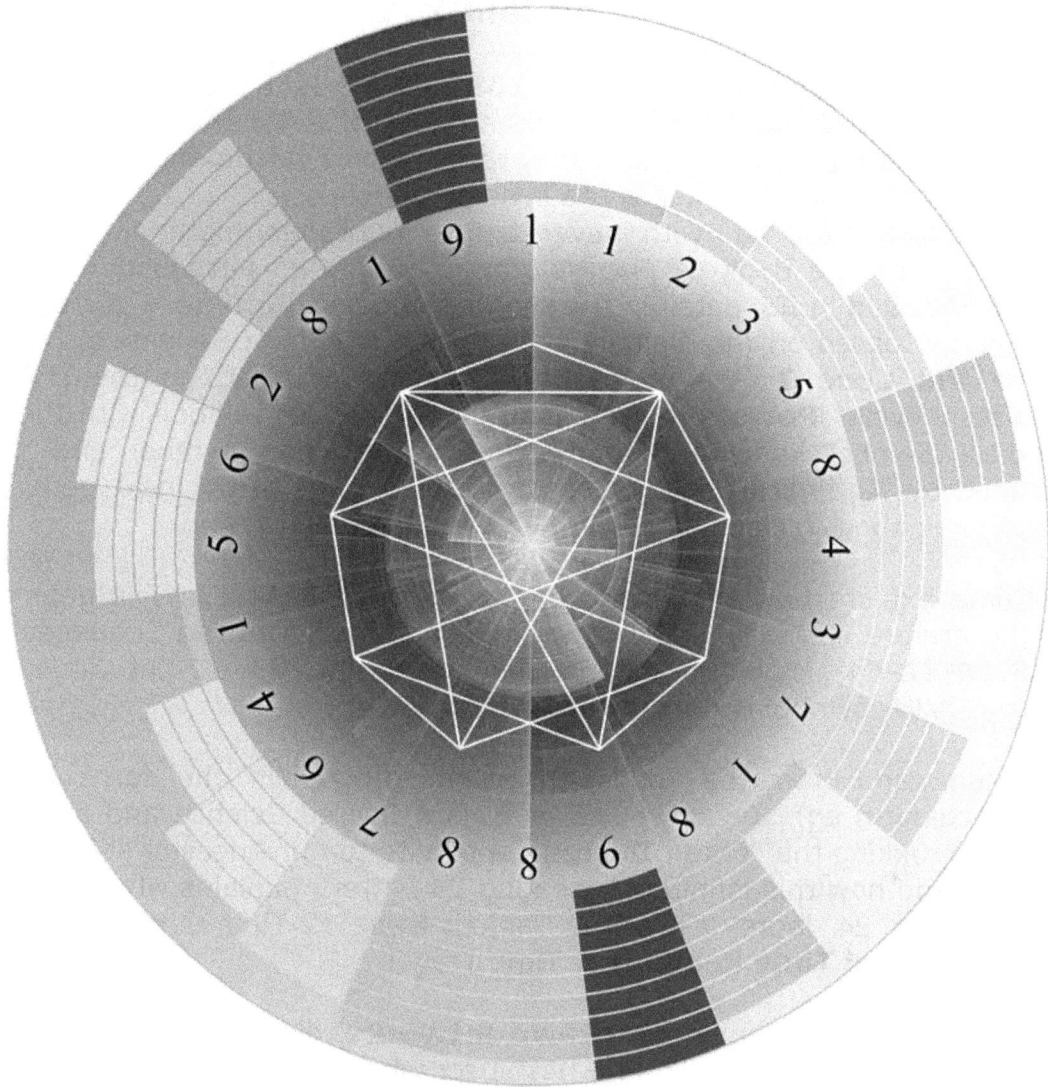

Phi Code 1 Psyho-Active Disk, timeless Wheel of 24
Mystic cogged wheels,
designed conceptually by Jain 108.
It can be used as a mat for charging glasses of water and drinks,
for energizing plates of food
for placing over injured parts of the body
etc
The Disks for Phi Codes 2 and 3 will be released soon.
(Computer Generated Artwork by Aurel Pumayana.

BACK COVER BLURB:
THE BOOK OF PHI, Volume 3
The 108 Codes, an Introduction.
By Jain 108

• What is all the fuss about this "**108 CODE**"! Is it the Pulse, the Rhythm, **the Living Curvation**, the True Mathematics of Nature? Is it the Essence of that which is repeatable, **shareable**, tangible and yet timeless, a phenomenal mathematical code that makes the invisible visible?

Such a rare diamond, with never before published material. Such a vast subject, that The Book Of Phi, Volume 3 is only an introduction to the 108 CODE revealed in full detail in the next two books: **Volume 4** is on **Phi Code 1** (linearly based on the digitally compressed Fibonacci Sequence), and **Vol 5** is on **Phi Code 2** (the multi-dimensional form based on the cryptic and endlessly cyclic "**Powers of Phi**" and has a direct correlation with the ancient Medical Symbol known as the winged **Cadduceus** symbol and currently claimed by big Pharma. The release of these codes to the next generation of children is about reclaiming our Power and Love).

• Come explore how the Divine Proportion (**1:1.618033...**) is geometrically and therefore eternally linked to the Binary Code (Doubling Sequence of **1-2-4-8-16-32-64**). When the data is compressed onto the 9 Point Circle, the well known "**VW**" symbol is **mathematically derived**!

• Understand how the forbidden **Numerology** (in the form of **Digital Compression** eg: 108 = 1+0+8 = 9) is a Sacred Science and an important Key to grokking the "**Fixed Design**", a tribute to our Ancestor's' **Law Of One**. Understand how this infinitely repeating 24 Pattern explains why we still divide our Day into 24 Hours! It is a true Galactic **Base 12** Time Code hinting clues to the **Physics of Time Bending** or Time Travel.

• With lots of beautiful **hand drawn artworks**, diagrams and worksheets to construct the Phi Spiral etc Jain has successfully taken the Lost Knowledge of the Phi Code 108 and highlighted its educational value.

• A billion Indians are worshipping this frequency of 108 yet they know not why. Jain, a full-blooded Phoenician, reveals his insights into Sri 108, becoming the cynosure of every Indian mathematician's eye. How remarkable, though simple, that the compression or digital reduction of the Fibonacci Sequence reveals a **distinct 24 Repeating Pattern** whose **sum adds up to 108**, not only one code, but 3. The patterns were always there, but we could not see them for thousands of years, as we were not trained to be engaged in the **visual feminine right brain mathematics**.

• Explore the 9 Point Circle, Wheels of 24 and much more.

tHE eND